Lecture Notes in Computer Science 10579

Commenced Publication in 1973
Founding and Former Series Editors:
Gerhard Goos, Juris Hartmanis, and Jan van Leeuwen

Editorial Board

More information about this series at http://www.springer.com/series/7409

Marieke van Erp et al. (Eds.)

Knowledge Graphs and Language Technology

ISWC 2016 International Workshops: KEKI and NLP&DBpedia
Kobe, Japan, October 17–21, 2016
Revised Selected Papers

Springer

Editors

see next page

ISSN 0302-9743   ISSN 1611-3349 (electronic)
Lecture Notes in Computer Science
ISBN 978-3-319-68722-3   ISBN 978-3-319-68723-0 (eBook)
https://doi.org/10.1007/978-3-319-68723-0

Library of Congress Control Number: 2017956739

LNCS Sublibrary: SL3 – Information Systems and Applications, incl. Internet/Web, and HCI

Printed on acid-free paper

This Springer imprint is published by Springer Nature
The registered company is Springer International Publishing AG
The registered company address is: Gewerbestrasse 11, 6330 Cham, Switzerland

Workshop Editors

Marieke van Erp
Digital Humanities Group
KNAW Humanities Cluster
Amsterdam
The Netherlands

Sebastian Hellmann (iD)
University of Leipzig
Leipzig
Germany

John P. McCrae (iD)
National University of Ireland
Galway
Ireland

Christian Chiarcos
Institut für Informatik
Goethe University Frankfurt
Frankfurt, Hessen
Germany

Key-Sun Choi
Division of Web Science and Technology,
   Department of Computer Science
KAIST
Daejeon
Korea (Republic of)

Jorge Gracia (iD)
Universidad Politécnica de Madrid
Madrid, Spain

Yoshihiko Hayashi
Waseda University
Tokyo
Japan

Seiji Koide
Ontolonomy LLC
Yokohama
Japan

Pablo Mendes
Apple San Francisco
San Francisco, CA
USA

Heiko Paulheim
Inst für Info & Wirtschaftsinfo
Universität Mannheim
Mannheim, Baden-Württemberg
Germany

Hideaki Takeda
National Institute of Informatics
Tokyo, Japan

Preface

This volume contains the combined proceedings of the papers presented at the First Workshop on Knowledge Extraction and Knowledge Integration (KÉKI 2016) and the 4th NLP&DBpedia workshop. Both workshops were held in conjunction with the 15th International Semantic Web Conference in Kobe, Japan, in October 2016.

The main focus of the KÉKI workshop is the use of linguistic linked open data. As more and more linguistic linked open data sources are becoming available (for example, through the linguistic linked open data or LLOD cloud) it is now time to start thinking and building linked data-aware natural language processing (NLP) applications. The focus of the NLP&DBpedia workshop is complementary to this as it focused on the linguistic aspects of DBpedia, one of the most popular openly available multilingual knowledge bases. The two subgoals of the NLP&DBpedia workshop are to improve DBpedia through NLP applications, and to boost NLP application by integrating knowledge from DBpedia. The KÉKI workshop received 11 submissions of which five were accepted for presentation and publication in this volume. The NLP&DBpedia workshop received nine submissions, of which four were accepted and presented. All papers were reviewed by at least three Program Committee members. Together, the nine submissions you find in this volume cover a broad spectrum of research on the intersection of knowledge graphs and language technology.

Roughly three dimensions can be discerned in the workshop submissions: (1) domain applications (Niekler and Kahmann; Evain, Vervaegen, and Matton), (2) quality assessment and improvement of knowledge graphs (Ell, Hakimov, and Cimiano; McCrae and Prangnawarat; Kim and Choi; and Yoshioka) and (3) use of knowledge graphs in NLP applications (Sasaski and Dojchinovski; Burget; and Van Erp and Vossen).

Furthermore, this volume also contains a contribution from Gerard de Melo (Rutgers University), the invited speaker from the NLP&DBpedia workshop, titled "Knowledge Graphs: Venturing out into the Wild."

We would like to thank the authors for their contributions and the Program Committee members for their reviews. We hope to see many more collaborations on the intersection of knowledge graphs and language technology.

June 2017

Marieke van Erp
Sebastian Hellmann
John P. McCrae
Christian Chiarcos
Key-sun Choi
Jorge Gracia
Yoshihiko Hayashi
Seiji Koide
Pablo Mendes
Heiko Paulheim
Hideaki Takeda

Organization

Program Committee

Sören Auer	University of Bonn, Germany
Caroline Barriere	CRIM, Canada
Christian Bizer	University of Mannheim, Germany
Georgeta Bordea	DERI Galway, National University of Ireland
Carmen Brando	Ecole des Hautes Etudes en Sciences Sociales
Volha Bryl	Springer Nature, Germany
Paul Buitelaar	Insight - National University of Ireland, Galway, Ireland
Philipp Cimiano	Bielefeld University, Germany
Thierry Declerck	DFKI GmbH, Germany
Dongpo Deng	Institute of Information Science, Academia Sinica, Taiwan
Richard Eckart de Castilho	Technische Universität Darmstadt, Germany
Agata Filipowska	Poznan University, Poland
Francesca Frontini	Paul-Valéry University, France
Anja Jentzsch	Hasso Plattner Institute, Germany
Bettina Klimek	Leipzig University, Germany
Dimitris Kontokostas	Leipzig University, Germany
Monica Monachini	Consiglio Nazionale delle Ricerche
Steven Moran	University of Zürich, Switzerland
Luis Morgado Da Costa	Nanyang Technological University, Singapore
Andrea Moro	Sapienza, Università di Roma, Italy
Yohei Murakami	Kyoto University, Japan
Roberto Navigli	Sapienza, Università di Roma, Italy
Naoaki Okazaki	Tohoku University, Japan
Wim Peters	University of Sheffield, UK
Mariano Rico	Universidad Politécnica de Madrid, Spain
Giuseppe Rizzo	ISMB, Italy
Víctor Rodriguez-Doncel	Universidad Politécnica de Madrid, Spain
Harald Sack	Hasso Plattner Institute, Germany
Felix Sasaki	DFKI/W3C, Germany
Michael Schuhmacher	Mannheim University, Germany
Christina Unger	CITEC, Universität Bielefeld, Germany
Ricardo Usbeck	Leipzig University, Germany
Marieke van Erp	KNAW Humanities Cluster, The Netherlands
Sebastian Walter	Semalytix GmbH, Germany
Ulli Waltinger	Siemens AG - Corporate Technology
Eveline Wandl-Vogt	Austrian Academy of Science, Austria

Haofen Wang   East China University of Science and Technology, China
Krzysztof Węcel   Poznan University of Economics, Poland
Masaharu Yoshioka   Hokkaido University, Japan

Contents

Knowledge Graphs: Venturing Out into the Wild

Gerard de Melo[✉]

Rutgers University, New Brunswick, NJ, USA
gdm@demelo.org
http://gerard.demelo.org

Abstract. While we now have vast collections of knowledge at our disposal, it appears that our systems often need further kinds of knowledge that are still missing in most knowledge graphs. This paper argues that we need keep moving further beyond simple collections of encyclopedic facts. Three key directions are (1) aiming at more tightly integrated knowledge, (2) distilling knowledge from text and other unstructured data, and (3) moving towards cognitive and neural approaches to better exploit the available knowledge in intelligent applications.

Keywords: Knowledge graphs · Information extraction · Neural methods

1 Introduction

In the past decade, knowledge graphs have grown from niche academic endeavours to becoming crucial assets for many IT companies. Well-known examples include DBpedia [13], YAGO [11], the Google Knowledge Graph, and Microsoft's Satori. Yet, although we now have vast repositories of facts at our disposal, it appears that our systems often need further kinds of knowledge that are still missing in most knowledge graphs.

This paper surveys three key directions to address the shortcomings of current large-scale knowledge graphs, suggesting paths for moving further beyond simple collections of encyclopedic facts. Section 2 focuses on better knowledge integration for structured data. Section 3 discusses how to connect structured data to the vast amounts of knowledge effectively locked away in unstructured sources. Finally, Sect. 4 proposes cognitive and neural approaches as a means of making better use of such knowledge.

2 Knowledge Integration for Structured Data

In the past, the knowledge acquisition bottleneck was often cited as a key challenge for artificial intelligence. Nowadays, there is a deluge of new sources of machine-readable knowledge. These include not only the RDF-based ones in the Linked Data cloud, but also thousands of open datasets stored in various other formats, and millions of web pages that incorporate structured data.

© Springer International Publishing AG 2017
M. van Erp et al. (Eds.): ISWC 2016 Workshops, LNCS 10579, pp. 1–9, 2017.
https://doi.org/10.1007/978-3-319-68723-0_1

While this abundance of different sources is certainly a blessing, it also brings a set of challenges in downstream applications wishing to make use of such data. What we have at our disposition is in several respects like a rich library with thousands of books. While this library may ultimately be able to serve our information needs, it is not always trivial to find relevant books and locate the desired facts within them. A very early pioneering attempt at addressing this, going even beyond individual libraries, was made in 1895 by Paul Otlet and Henri Lafontaine in their Répertoire Bibliographique Universel (RBU). This universal index would eventually grow to over 15 million index cards, aspiring to systematically organize much of the world's knowledge.

In the digital age, we need tools and algorithms that provide a similar level of universal knowledge organization, yielding pertinent data for a given information need. Converting the various input data formats to a common form such as RDF is just the first step. A more significant challenge is overcoming the heterogeneity of their data models and their incongruent forms of knowledge organization.

One aspect is connecting entities across datasets. The simplest case is when there are shared identifiers. For instance, many resources are linked to Wikipedia or DBpedia for general entities, Lexvo.org [19] for linguistic entities, and Word-Net for sense identifiers. In general, however, creating links remains very challenging, despite the long history of work on this. This is particularly true when we aim at entity matching not just between two sources but across a large range of datasets, as this is best done jointly so as to exploit the mutual influence between various candidate matches. The LINDA approach [3] addresses this via a scalable greedy approach that first establishes those links that appear to be easy and reliable. Information about these accepted links is then used to update our beliefs about the accuracy of other potential links. Further algorithms allow us to check for the consistency of entity match links [18]. Another little-studied but important problem is the issue of varying levels of granularity of concepts [21]. Even an entity name such as "London" may refer to multiple competing notions of the entity, e.g., the small City of London, the London metropolitan area, Greater London, the Greater London Built-up Area, or others that may extend as far as to include London Gatwick airport, in addition to various historic definitions. Establishing entity-level links allows us to connect various resources in a cloud of linked resources, similar to the general Linked Data cloud, but possibly also for specific domains as in the Linguistic Linked Data cloud [17].

Even with such links, however, the knowledge is not fully integrated. We have developed algorithms that take a series of separate knowledge graphs as input and produce a single coherent taxonomy, based on ontological principles [1,26].

Another important step is to connect the various properties that are in use across different datasets. To this end, the FrameBase project provides a large schema [28] based on verbs in the English language. This schema draws on the FrameNet lexical resource, extended with additional entries from WordNet for greater coverage. Within the project, a number of heuristics have been developed to automatically connect other ontologies and vocabularies to FrameBase [31]. In some cases, however, manual modeling may be necessary to extend Frame-Base to cover more specific properties that cannot straightforwardly be aligned with FrameBase via a 1-to-1 mapping [30]. Hence, we have also developed a user interface that facilitates engaging human experts to define more complex mappings [32].

Examples of integrated knowledge graphs include Lexvo.org [19], which describes languages, scripts, words, and other language-related units, the Universal Wordnet (UWN) [25], which provides multilingual word meanings and their relationships, and MENTA [26], a multilingual taxonomy coherently combining over 200 language editions of Wikipedia. Open challenges include how to cope with incompatible licenses. For instance, the Open PHACTS portal provides data from different sources with incompatible licenses, some of which do not permit derivatives.

3 Connecting Unstructured Data

While information systems excel at processing structured data, large amounts of the world's knowledge are only available via other modalities.

3.1 Text and Language

For natural language text, suitable methods are needed for analysis and knowledge extraction. Standard forms of information extraction (IE) consider only a narrow, predefined set of relations. Although there has been significant progress in this area, including drawing on Web-scale data [36], relying on entire knowledge graphs as seed data [39], and using deep learning models [42], their success often hinges on the availability of relevant training or seed data for each relation.

Open information extraction is a well-known alternative, aiming to cover arbitrary relationships encountered in a text. This open-ended approach may support a broader range of applications. For instance, the PEAK system [46] shows how this allows us to automatically evaluate the quality of a textual summary, given reference summaries. While measures such as ROUGE are often used to automatically evaluate text summarization systems, ROUGE only works reliably when averaging across numerous different texts to be summarized. Often, however, we only wish to evaluate a single summary. This might be a student-written one, for instance, used as a means of assessing reading comprehension. PEAK fills this gap using open IE: Subject-predicate-object triples are used to discover salient units of content expressed in a summary and then such units can be compared between a student-written summary and high-quality reference summaries to automatically assess the student's reading comprehension.

Still, open IE is perhaps best used only as an internal component of knowledge-driven systems. Although the extractions are very useful in certain tasks, they are not sufficiently clean and normalized to be similar to what one encounters in curated knowledge graphs. Additionally, they also mostly neglect n-ary relations (for $n > 2$).

Instead, it may be desirable to obtain extractions with a more well-defined target representation for the extracted knowledge. In particular, is often beneficial for such representations to be compatible with those used in structured knowledge graphs. There is a growing number of knowledge graphs capturing linguistic information. Examples include the aforementioned Lexvo.org [19] and Universal Wordnet projects [25, 26]. However, one needs a wide-coverage schema that also covers most kinds of relationships that one expects to extract from text. To this end, the aforementioned FrameBase schema is a suitable target [28], as it brings English verbs and their arguments into the realm of Linked Data. By drawing on interlinked resources such as the Universal WordNet [25], languages other than English are also connected to it. Third-party tools such as the PIKES [6] and KNEWS [2] systems can take us from raw text to extractions based on the FrameBase schema.

Apart from improving the overall accuracy of such systems, ongoing research is focusing on coping with the various intricacies of natural language. Particular phenomena that are being worked on include ambiguity [38], metaphor [33], comparisons [40], nominalizations [8, 27], and abstract events [29]. Of particular importance to knowledge-driven applications is the status of clauses and phrases. When a text discusses the "dismissal of the Ambassador", then in some, but not all contexts, we can conclude that the Ambassador has been dismissed. If a text states that everyone "dislikes that the company is releasing new product X" then the machine should be able to infer that they are indeed releasing X, whereas if it states that they "deny that the company is releasing new product X", then it is not clear. Similarly, "refusing to secure a loan" is quite different from "managing to secure a loan". Although these phrases seem trivial for humans to interpret, they differ merely in individual words, and information extraction systems ought to be able to make sense of these differences. We have developed a prototype system that achieves this [23].

Finally, apart from going from text to knowledge, there are also further tasks at the intersection of language and knowledge. An obvious one is to consider the inverse direction of going from knowledge to text, i.e., text generation [43]. Another important task is to make knowledge searchable, i.e., retrieving facts as answers to a natural language query [14, 22]. All of these tasks relate to the current trend of developing intelligent conversational agents that rely on natural language skills as well as on knowledge.

3.2   Multimodal Knowledge

In recent years, computer vision has made significant progress and multimodal data has become more connected to language and knowledge. Large cultural heritage collections have become available as Linked Data. Standard computer vision datasets such as ImageNet and Visual Genome are connected to WordNet. Moreover, natural language captions can now be generated automatically for both images and video, by combining deep convolutional neural networks with recurrent models, perhaps incorporating ideas such as multi-faceted attention [15].

Ongoing research is targeting how to go beyond object detection and classification to gain a more complete and thorough understanding of what is going on in an image or video. One direction is to understand images at a higher level of abstraction by predicting not just the concrete objects that they portray, but also the overall activity [9]. This is challenging, because an activity such as *playing a game* may appear in countless different ways in an image or video, depending on the kind of game, the environment, the players, and the type of recording. The Knowlywood knowledge base collects large amounts of activity knowledge and images from Hollywood movies, among other sources [37]. Conversely, statistics from large image and video collections can also improve natural language processing [33].

Another direction is to aim at more fine-grained knowledge, by not just classifying rectangular bounding boxes, but obtaining a detailed pixel-level analysis of shapes and contours. In fact, our ShapeLearner project [44] takes this one step further and provides pixel-level information about the parts of an object, e.g. distinguishing an animal's head from the rest of its body, or distinguishing the grip of a sword from its blade. ShapeLearner is thus both a knowledge graph and an image analysis engine.

User interfaces also greatly benefit from multimodality. Knowledge base entities can be visualized both temporally and geographically [10, 11]. Queries may be multimodal as well. In a recent paper, we provided the first major steps towards multimodal question answering over Linked Data [14]. As mobile usage prevails, people now often have information needs that pertain to their surroundings and are best captured using an image.

4   Towards Cognitive and Neural Approaches

4.1   Neural Models

The ultimate goal of most knowledge bases is to enable more informed and intelligent applications. It has long been obvious that this will often require forms of inference that go beyond formal logical reasoning. For example, extracted and inferred knowledge assertions often come with confidence scores or probabilities that ought to be considered. In recent years, deep learning and other neural approaches have shown significant promise in this regard, enabling effective data-driven learning and inference for tasks that had just a few years go appeared intractable.

One important direction is to study semantic representations using neural methods. While well-known methods such as word2vec exploit the co-occurrence of words in large monolingual text corpora, recent work shows how to go beyond them and exploit further available cues in the data. One approach is to draw on information extraction to obtain higher-quality word embeddings [4]. We can also exploit document labels to learn high-quality representations for domain-specific concepts such as "carboplatin" or "prenatal exposure delayed effects" [16]. Additionally, it is now possible to obtain massively multilingual word representations covering many different languages simultaneously [7]. Last but not least, we can draw on large knowledge bases to learn embeddings for millions of entities in different languages [20].

Another direction is to investigate knowledge-driven applications of such representations in deep neural architectures. Currently, deep learning approaches are being investigated to discover salient information in text [45] and for neural information retrieval and ranking [12].

4.2 Common-Sense Knowledge

The final frontier is to go beyond learning towards genuinely intelligent behavior. This involves collecting substantial amounts of common-sense knowledge, which can take a number of different forms. We have investigated mining large amounts of basic commonsense knowledge assertions [39], fine-grained attributes [41], comparative commonsense knowledge [40] (e.g., that a falcon is faster than a leopard), and activity knowledge [37]. Such commonsense knowledge has been shown to aid in particularly challenging AI tasks such as metaphor interpretation [33].

However, human knowledge is unbounded and it is hence not sufficient to simply collect commonsense knowledge facts. For additional inference, we have investigated axiomatic rules [24] and large-scale reasoning [34,35]. Our latest approach is to combine commonsense knowledge with neural knowledge modeling, as exemplified by our WebBrain system [5]. WebBrain learns a neural model both from commonsense knowledge acquired from the Web as well as from general semantics as captured in word vector representations. With this knowledge, it attempts to make educated guesses beyond what has been observed on the Web. For example, WebBrain may guess that cockatiels are likely capable of flying, based on their similarity to other kinds of birds that fly.

5   Conclusion

While knowledge graphs have become widespread in industry and academia, we have seen that simple forms of structured facts do not fully resolve the traditional knowledge acquisition bottleneck.

In the future, systems will need to jointly learn and reason with knowledge from multiple heterogeneous sources of Big Data, including knowledge extracted from text, media, and large-scale structured knowledge repositories.

References

1. Bansal, M., Burkett, D., de Melo, G., Klein, D.: Structured learning for taxonomy induction with belief propagation. In: Proceedings of ACL 2014 (2014)
2. Basile, V., Cabrio, E., Schon, C.: KNEWS: Using logical and lexical semantics to extract knowledge from natural language. In: Proceedings of ECAI (2016)
3. Böhm, C., de Melo, G., Naumann, F., Weikum, G.: LINDA: distributed web-of-data-scale entity matching. In: Proceedings of CIKM 2012. ACM (2012)
4. Chen, J., Tandon, N., de Melo, G.: Neural word representations from large-scale commonsense knowledge. In: Proceedings of WI 2015 (2015)

5. Chen, J., Tandon, N., Hariman, C.D., de Melo, G.: WebBrain: joint neural learning of large-scale commonsense knowledge. In: Groth, P., Simperl, E., Gray, A., Sabou, M., Krötzsch, M., Lecue, F., Flöck, F., Gil, Y. (eds.) ISWC 2016. LNCS, vol. 9981, pp. 102–118. Springer, Cham (2016). doi:10.1007/978-3-319-46523-4_7

6. Corcoglioniti, F., Rospocher, M., Palmero Aprosio, A.: Frame-based ontology population with PIKES. TKDE **28**(12), 3261–3275 (2016)

7. de Melo, G.: Wiktionary-based word embeddings. In: Proceedings of MT Summit XV (2015)

8. Freitas, C., de Paiva, V., Rademaker, A., de Melo, G., Real, L., Silva, A.: Extending a lexicon of Portuguese nominalizations with data from corpora. In: Baptista, J., Mamede, N., Candeias, S., Paraboni, I., Pardo, T.A.S., Volpe Nunes, M.G. (eds.) PROPOR 2014. LNCS, vol. 8775, pp. 114–124. Springer, Cham (2014). doi:10.1007/978-3-319-09761-9_12

9. Gan, C., Lin, M., Yang, Y., de Melo, G., Hauptmann, A.G.: Concepts not alone: exploring pairwise relationships for zero-shot video activity recognition. In: Proceedings of AAAI. AAAI Press (2016)

10. Ge, T., Wang, Y., de Melo, G., Li, H.: Visualizing and curating knowledge graphs over time and space. In: Proceedings of ACL 2016. ACL (2016)

11. Hoffart, J., Suchanek, F.M., Berberich, K., Lewis-Kelham, E., de Melo, G., Weikum, G.: YAGO2: exploring and querying world knowledge in time, space, context, and many languages. In: Proceedings of WWW 2011. ACM (2011)

12. Hui, K., Yates, A., Berberich, K., de Melo, G.: A position-aware deep model for relevance matching in information retrieval. CoRR abs/1704.03940 (2017). http://arxiv.org/abs/1704.03940

13. Lehmann, J., Isele, R., Jakob, M., Jentzsch, A., Kontokostas, D., Mendes, P., Hellmann, S., Morsey, M., van Kleef, P., Auer, S., Bizer, C.: DBpedia - a large-scale, multilingual knowledge base extracted from Wikipedia. Semant. Web **6**(2), 167–195 (2015)

14. Li, H., Wang, Y., de Melo, G., Tu, C., Chen, B.: Multimodal question answering over structured data with ambiguous entities. In: Proceedings of WWW 2017 (2017)

15. Long, X., Gan, C., de Melo, G.: Video captioning with multi-faceted attention. CoRR abs/1612.00234 (2016). http://arxiv.org/abs/1612.00234

16. Loza Mencía, E., de Melo, G., Nam, J.: Medical concept embeddings via labeled background corpora. In: Proceedings of LREC 2016 (2016)

17. McCrae, J.P., Chiarcos, C., Bond, F., Cimiano, P., Declerck, T., de Melo, G., Gracia, J., Hellmann, S., Klimek, B., Moran, S., Osenova, P., Pareja-Lora, A., Pool, J.: The open linguistics working group: developing the linguistic linked open data cloud. In: Proceedings of LREC 2016 (2016)

18. de Melo, G.: Not quite the same: identity constraints for the Web of Linked Data. In: Proceedings of AAAI, pp. 1092–1098. AAAI Press (2013)

19. de Melo, G.: Lexvo.org: language-related information for the linguistic linked data cloud. Semantic Web **6**(4), 393–400 (2015)

20. de Melo, G.: Inducing conceptual embedding spaces from Wikipedia. In: Proceedings of WWW 2017. ACM (2017)

21. de Melo, G., Baker, C.F., Ide, N., Passonneau, R., Fellbaum, C.: Empirical comparisons of MASC word sense annotations. In: Proceedings of LREC 2012 (2012)

22. de Melo, G., Hose, K.: Searching the web of data. In: Serdyukov, P., Braslavski, P., Kuznetsov, S.O., Kamps, J., Rüger, S., Agichtein, E., Segalovich, I., Yilmaz, E. (eds.) ECIR 2013. LNCS, vol. 7814, pp. 869–873. Springer, Heidelberg (2013). doi:10.1007/978-3-642-36973-5_105

23. de Melo, G., de Paiva, V.: Sense-specific implicative commitments. In: Gelbukh, A. (ed.) CICLing 2014. LNCS, vol. 8403, pp. 391–402. Springer, Heidelberg (2014). doi:10.1007/978-3-642-54906-9_32

24. de Melo, G., Suchanek, F., Pease, A.: Integrating YAGO into the suggested upper merged ontology. In: Proceedings of ICTAI 2008. IEEE Computer Society (2008)

25. de Melo, G., Weikum, G.: Towards universal multilingual knowledge bases. In: Proceedings of the 5th Global WordNet Conference, pp. 149–156 (2010)

26. de Melo, G., Weikum, G.: Taxonomic data integration from multilingual Wikipedia editions. Knowl. Inf. Syst. **39**(1), 1–39 (2014)

27. de Paiva, V., Real, L., Rademaker, A., de Melo, G.: NomLex-PT: a lexicon of Portuguese nominalizations. In: Proceedings of LREC 2014. ELRA, May 2014

28. Rouces, J., de Melo, G., Hose, K.: FrameBase: representing N-Ary relations using semantic frames. In: Gandon, F., Sabou, M., Sack, H., d'Amato, C., Cudré-Mauroux, P., Zimmermann, A. (eds.) ESWC 2015. LNCS, vol. 9088, pp. 505–521. Springer, Cham (2015). doi:10.1007/978-3-319-18818-8_31

29. Rouces, J., de Melo, G., Hose, K.: Representing specialized events with FrameBase. In: Proceedings of the 4th International Workshop on Detection, Representation, and Exploitation of Events in the Semantic Web (DeRiVE) at ESWC 2015 (2015)

30. Rouces, J., de Melo, G., Hose, K.: Complex schema mapping and linking data: beyond binary predicates. In: Proceedings of LDOW 2016 (2016)

31. Rouces, J., de Melo, G., Hose, K.: Heuristics for connecting heterogeneous knowledge via FrameBase. In: Sack, H., Blomqvist, E., d'Aquin, M., Ghidini, C., Ponzetto, S.P., Lange, C. (eds.) ESWC 2016. LNCS, vol. 9678, pp. 20–35. Springer, Cham (2016). doi:10.1007/978-3-319-34129-3_2

32. Rouces, J., de Melo, G., Hose, K.: Klint: Assisting integration of heterogeneous knowledge. In: Proceedings of IJCAI 2016 (2016)

33. Shutova, E., Tandon, N., de Melo, G.: Perceptually grounded selectional preferences. Proceedings of ACL **2015**, 950–960 (2015)

34. Suda, M., Sutcliffe, G., Wischnewski, P., Lamotte-Schubert, M., de Melo, G.: External sources of axioms in automated theorem proving. In: Mertsching, B., Hund, M., Aziz, Z. (eds.) KI 2009. LNCS, vol. 5803, pp. 281–288. Springer, Heidelberg (2009). doi:10.1007/978-3-642-04617-9_36

35. Sutcliffe, G., Suda, M., Teyssandier, A., Dellis, N., de Melo, G.: Progress towards effective automated reasoning with world knowledge. In: Proceedings of the 23rd International FLAIRS Conference, pp. 110–115. AAAI Press (2010)

36. Tandon, N., de Melo, G.: Information extraction from web-scale n-gram data. In: SIGIR 2010 Web N-gram Workshop, vol. 5803, pp. 8–15. ACM (2010)

37. Tandon, N., de Melo, G., De, A., Weikum, G.: Knowlywood: mining activity knowledge from Hollywood narratives. In: Proceedings of CIKM 2015 (2015)

38. Tandon, N., de Melo, G., Suchanek, F.M., Weikum, G.: WebChild: harvesting and organizing commonsense knowledge from the web. In: Proceedings of WSDM. ACM (2014)

39. Tandon, N., de Melo, G., Weikum, G.: Deriving a Web-scale common sense fact database. In: Proceedings of AAAI, pp. 152–157. AAAI Press (2011)

40. Tandon, N., de Melo, G., Weikum, G.: Acquiring comparative commonsense knowledge from the web. In: Proceedings of AAAI, pp. 166–172. AAAI (2014)

41. Tandon, N., de Melo, G., Weikum, G.: WebChild 2.0: fine-grained commonsense knowledge distillation. In: Proceedings of ACL 2017. ACL (2017)

42. Wang, L., Cao, Z., de Melo, G., Liu, Z.: Relation classification via multi-level attention CNNs. In: Proceedings of ACL 2016 (2016)

43. Wang, Y., Ren, Z., Theobald, M., Dylla, M., de Melo, G.: Summary generation for temporal extractions. In: Hartmann, S., Ma, H. (eds.) DEXA 2016. LNCS, vol. 9827, pp. 370–386. Springer, Cham (2016). doi:10.1007/978-3-319-44403-1_23
44. Xu, H., Wang, Y., Feng, K., de Melo, G., Wu, W., Sharf, A., Chen, B.: Shape-learner: towards shape-based visual knowledge harvesting. In: Proceedings of ECAI 2016, pp. 435–443. IOS Press (2016)
45. Yang, Q., Cheng, Y., Wang, S., de Melo, G.: HiText: text reading with dynamic salience marking. In: Proceedings of WWW 2017. ACM (2017)
46. Yang, Q., Passonneau, R.J., de Melo, G.: PEAK: Pyramid evaluation via automated knowledge extraction. In: Proceedings of AAAI. AAAI Press (2016)

Information Extraction from the Web by Matching Visual Presentation Patterns

Radek Burget(✉)

Faculty of Information Technology, Centre of Excellence IT4Innovations,
Brno University of Technology, Bozetechova 2, 612 66 Brno, Czech Republic
`burgetr@fit.vutbr.cz`

Abstract. The documents available in the World Wide Web contain large amounts of information presented in tables, lists or other visually regular structures. The published information is however usually not annotated explicitly or implicitly and its interpretation is left on a human reader. This makes the information extraction from web documents a challenging problem. Most existing approaches are based on a top-down approach that proceeds from the larger page regions to individual data records, which depends on different heuristics. We present an opposite bottom-up approach. We roughly identify the smallest data fields in the document and later, we refine this approximation by matching the discovered visual presentation patterns with the expected semantic structure of the extracted information. This approach allows to efficiently extract structured data from heterogeneous documents without any kind of additional annotations as we demonstrate experimentally on various application domains.

Keywords: Web data integration · Information extraction · Structured record extraction · Page segmentation · Content classification · Ontology mapping

1 Introduction

The World Wide Web contains a vast amount of documents containing data records presented in a regular, visually consistent way using different kinds of lists, tables or other logical structures. Typical examples include product data, events, exchange rates, sports results, timetables and many more. Although the structure of the presented information is generally predictable for every application domain, the actual data records may be presented in the HTML documents in countless ways.

For large and consistent data sources such as Wikipedia, it is possible to define extraction templates that may be reused for a great number of pages. However, for heterogeneous sources where every document may use different presentation patterns, this approach is not feasible. The great variability in presentation and almost no semantic annotations available in HTML documents

© Springer International Publishing AG 2017
M. van Erp et al. (Eds.): ISWC 2016 Workshops, LNCS 10579, pp. 10–26, 2017.
https://doi.org/10.1007/978-3-319-68723-0_2

make the automatic integration of such web sources to structured datasets (such as DBPedia) a challenging problem.

In this paper, we present a method for the discovery and extraction of structured records in web documents. In contrast to most current approaches that perform a complex analysis of the document HTML code or its visual organization in order to detect repeating structures (top-down approach) [1,8,13,14], we use an opposite (bottom-up) approach: We start with the smallest consistent text elements and we match the visual relationships among these elements with the expected structure of the extracted records. This way, we are able to automatically discover the visual patterns used for presenting the data records in the given document.

The most important benefits of the presented approach are the following:

- The extraction task specification is based only on a generic domain knowledge consisting of the logical relationships among the individual data fields to be extracted and a very general specification of allowed values for each data field.
- No templates need to be used and no labels or annotations are required in the source documents.
- The method can be easily adapted for any target domain as it allows integration of arbitrary domain-specific knowledge (such as dictionaries or extracted data formats) and different data field recognition methods (from domain-specific heuristics to general NLP methods such as named entity recognition). We demonstrate the method application to different target domains in Sect. 9.
- The method is independent on the format of the input documents. We use the HTML and PDF documents as the most important information source but any other document type where the styled text is available may be used as well.

We also demonstrate that our information extraction method may be integrated with DBPedia in two ways: (1) DBPedia may be used for the recognition of candidate data fields in the extracted records and (2) the extracted records may contain new data that may be linked back to existing DBPedia resources. This allows integrating new web sources to DBPedia.

2   Related Work

Information extraction from web documents is a research area that is interesting for different applications. The most important application areas include extracting data results from query result pages [1,8,9,12–15] (obtained either from general search engines or specialized ones such as product search) or obtaining structured data buried in large sets of web documents [5,10].

When considering the recently published approaches, we may identify two basic groups from the perspective of the used representation of the input document: (1) code-based approaches that use a representation of the input document code (mainly DOM for HTML documents) [6,9,10,15] and (2) vision-based approaches that use some kind of visual representation of the rendered page that may be obtained by adding some visual features to the document code model [1,8] or by using a standard page segmentation algorithm [13,14]. However, regardless of the used document representation, all the mentioned approaches expect HTML documents at the input.

Most existing methods are based on a top-down approach which is basically presentation-driven. After creating the document model as mentioned above, the model is usually preprocessed in order to filter the content blocks regarded as noise or to locate the most probable regions of interest (called a result section [12], data sections [14] or data region [8]). Then, the individual data records are identified based on the detection of repeating structures in the model by frequency measures [9] or visual pattern detection [1,12,14]. The structure of the extracted information is inferred from the discovered records while using additional information such as explicit labels present in the page [1,12,14,15] or even the query interface in case of the query result extraction [12,13]. This presentation-driven approach is suitable for many applications such as the deep web crawling. On the other hand, in case of information extraction from web sources for the semantic web, structured databases or particular applications, the structure of the extracted information is typically available in advance (for example as a domain ontology) and the task is to locate the corresponding data records in the input documents.

We have identified only a few approaches that are based on a previously known ontological model of the information being extracted. The classical work by Embley et al. [6] uses a conceptual domain model that defines the lexical and non-lexical classes and the relationships among them. However, before the conceptual model may be used for information extraction, a complex input document preprocessing is required that does not take the into account the domain model and it is based on heuristics tightly related to the HTML language constructions. Similarly, our earlier work [2] uses complex vision-based document preprocessing for creating a logical model of the processed document in a form that can be later matched with an ontology-based domain model.

Our approach we present in this paper shares many ideas regarding the ontological specification of the target domain with the work of Embley et al. [6]. However, instead of a complicated document preprocessing that presents a potential point of failure, we attempt to use the ontological specification as early as possible. As we mention in the introduction, our approach proceeds in a bottom-up manner leaving the presentation style analysis to later stages. This allows to avoid the complex document preprocessing that is usually HTML-specific and it presents a potential source of errors.

We have successfully tested some of the presented concepts during the Sem-Pub 2015 challenge [5]. Our solution [11] was however tailored to a given particular application. In this paper, we present a new method based on a general model of the target domain.

3 Task Specification

The goal of our method is extracting information corresponding to ontological *concepts* (classes) from documents. In Fig. 1, we show a sample class (a conference paper) that is taken from a larger ontology we used for a particular information extraction task [11].

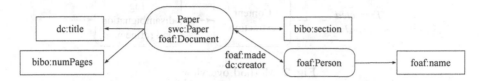

Fig. 1. Sample ontology representing a concept (Paper) and its data and object properties using the concepts and properties from the Bibliographic Ontology, FOAF Ontology and the Semantic Web Conference Ontology. The ovals represent the object properties and the rectangles represent its data properties.

According to the usual terminology in this area (for example [6]), the information about the instances of the given class (*individuals*) is represented by *data records* in the source documents. Each data record consists of multiple *data fields*, sometimes also called data units [13] that provide the lexical representation of some data properties (lexical properties) of the individual. The data fields are represented as text strings contained in the document text. Thus, a data record can be defined as a set of data fields that describe the same individual.

The task we investigate in this paper is to recognize all the data records in the source documents that belong to a single entity that is known in advance. Considering the Paper class in Fig. 1 as the input concept specification, the task is to recognize all the data records in the source documents that contain the information about individual papers containing their titles, author names, sections and pages.

4 Method Overview

We assume processing of web documents containing multiple data records corresponding to the same concept. The data records are presented in one or more source documents in a visually consistent way (we discuss the visual consistency in more detail in Sect. 7). The key idea is to discover the most frequent visual presentation patterns that occur in the source documents and that are used for presenting the data records. Subsequently, the data records are extracted using the discovered patterns. The method in general does not involve any learning phase on a training set of documents. For every extraction task, it only analyses the presentation patterns in the given source document. However, a trained

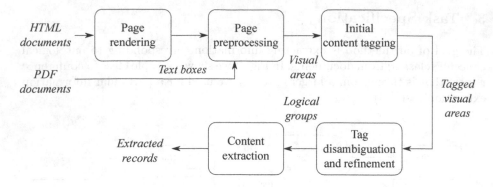

Fig. 2. Method overview

classifier may be used as one of the sources of the necessary background knowledge for certain application domains. We demonstrate one such application in Sect. 9.4.

Figure 2 shows the overview of our method. It operates on a visual representation of the source documents that is independent on the underlying code. Therefore, the first step is the document preprocessing that consists of creating an uniform representation of the source documents as a set of *visual areas*.

Next, in the initial tagging step, we perform an approximate recognition of the individual parts of the document content. This step gives a rough idea about the possible meaning of the individual visual areas; that means which visual areas might possibly correspond to some particular data fields. The result is represented by adding *tags* to the respective visual areas. Since the initial tagging is only approximate, some visual areas may obtain multiple tags and some of them may be tagged incorrectly.

Therefore, in the next step, we discover the most frequent presentation patterns used in the source documents and we use them to disambiguate and refine the assigned tags. The most supported visual patterns are then used for recognizing the desired data records.

In the following sections, we discuss the details of all the individual steps.

5   Input Document Preprocessing and Representation

The purpose of the input document preprocessing is to create a unified, format-independent model of the document content and its visual presentation. This step is typical for all the visually oriented information extraction approaches; we may mention the Visual block model used in [1], Page layout model [2] or the Visual block tree in [13].

Typically, all these models have a hierarchical structure which corresponds to the typical visual organization of the content in a web page. However, in our approach, we do not take into account the overall visual organization of the page such as visually separated block or sections. Instead, we employ a bottom-up approach, that considers only the individual parts of the text content, their visual style and mutual visually expressed relationships. Therefore, we do not need to represent the complete visual hierarchy of the page and we use only a simplified flat model consisting of a set of *visual areas* as we define below.

The input of the preprocessing step is a set of *text boxes* contained in the source document. With a text box, we understand a rectangular area in the displayed page with a know position, size containing a portion of the document text. For HTML documents, the information about the text boxes is available from a rendering engine after the document has been rendered. In case of PDF documents, this information is directly available in the source document. In both cases, the information about the visual style of the contained text (such as the used font or color) is also available for each box.

In the preprocessing step, we create visual areas from the text boxes. A visual area provides an abstraction over the rendered boxes. It is a rectangular area in the rendered page that corresponds to one or more displayed text boxes depending on the chosen *granularity* as we explain below. We define a visual area a as follows:

$$a = (rect, text, style, B) \tag{1}$$

where $rect = (x, y, w, h)$ is a rectangle representing the area position and size in the rendered page, $text$ is the text string contained in the area and $style$ is the area style:

$$style = (fs, w, st, c, bc) \tag{2}$$

where fs is the average font size used in the visual area, $w \in [0, 1]$ is the average font weight where 1 means the whole area written in bold font and 0 means the whole area written in regular font, $st \in [0, 1]$ is the average font style (1 for italic font, 0 for regular font) and c and bc are the foreground and background colors used in the area. Finally, $B = b_1, b_2, \ldots, b_n$ is the set of boxes contained in the area $(n > 0)$.

As the result of the preprocessing step, we obtain a set A of all the visual areas in the page:

$$A = a_1, a_2, \ldots, a_m \tag{3}$$

where m is the total number of visual areas in the page. Then, for any pair of visual areas $a_i, a_j \in A$ the corresponding sets of boxes B_i, B_j are disjoint $(B_i \cap B_j = \varnothing)$ and the corresponding rectangles $rect_i$, $rect_j$ do not overlap.

5.1 Visual Area Granularity

The granularity of the visual areas generally depends on the application domain. In Sect. 9, we give several examples of the application domains and we discuss the chosen visual area granularity for each of them. The highest possible granularity is obtained when for each visual area $a_i \in A$, the corresponding set of boxes B_i contains a single box ($|B_i| = 1$). However, for most application, we choose a higher granularity ($|B_i| \geq 1$). Typical choices are the following:

- *Inline-level granularity* – the visual areas are formed by sets of neighboring boxes (based on their positions on the page) that are vertically aligned to a single line and they share a consistent visual style as defined in (2). This level approximately corresponds to inline-level elements used in HTML documents.
- *Block-level granularity* – the visual areas are formed by sets of boxes that form a visually separated block of text in the page (for example a text paragraph). We use a simple block detection method proposed in [3] that is based on the discovery of clusters of adjacent boxes based on their positions in the page.

Depending on the chosen granularity, we obtain a larger or smaller set A of visual areas that represent the elementary pieces of the document content in the following steps of information extraction.

6   Initial Content Tagging

The purpose of the initial content tagging is to recognize all the visual areas that possibly might correspond to an extracted data field. Shortly, we want to identify the pieces of information that possibly "look like" some data field (for example a paper author name) when viewed separately. Each visual area is considered separately and it is assigned *tags* that indicate its possible meanings.

Based on the target domain, we define a set $T = t_1, t_2, \ldots, t_n$ of tags that may be assigned to visual areas. Each tag is identified by its name and it represents a particular data field to be extracted. For example, for the domain of conference papers shown in Fig. 1, we obtain the following set of tags corresponding to the data properties of the papers:

$$T = \{\text{title}, \text{authors}, \text{section}, \text{pages}\} \tag{4}$$

where the individual tags denote the paper title, author names, section title and page numbers respectively. For each tag, we define a *tagging* function that assigns a *support* to every visual area and tag:

$$tagging : A \times T \rightarrow \mathbb{R}_{[0,1]} \tag{5}$$

For a visual area $a \in A$ and a tag $t \in T$, the assigned support is a number $s \in [0, 1]$ that represents the probability that the visual area has the meaning that corresponds to the given tag. When $s > 0$, we say that the tag t has been assigned to a with the given support; for $s = 0$, we say that t has not been assigned to a. Multiple tags may be assigned to a single area (for example, the number *"15"* may be recognized as both hour and minutes in the time domain).

The initial tagging represents a highly approximate estimation of the meaning of the individual visual areas which is used as a starting point for further refining. We note that some of the tags (such as *title* and *section*) cannot be reliably distinguished when considering the visual areas separately. In that case, the visual area may obtain both tags (that means it may correspond to both the paper and section title) and later, the tags are disambiguated using the presentation context as described in Sect. 7.

From the practical point of view, we implement the *tagging* function as a set of *taggers* where the tagger is a procedure that is responsible for computing the support of a single particular tag given a visual area. The tagger implementation may be very variable but generally, we consider the following approaches to the tag assignment that may be combined arbitrarily:

- *DBPedia concept annotation* for example using the *Spotlight* tool [4].
- *Named entity recognition (NER)* may be used for recognizing the entities such as personal names or locations depending on the used NER classifier.
- Occurrences of *keywords* (for example month names), *numerical values* in given ranges or specific *regular expressions*.
- *Visual classification.* As we have shown in our earlier work [3], it is possible to use the visual features of the areas such as the used font, colors, position within the page or amount of contained text to create a classifier, that is first trained on a set of manually annotated documents and then, it may be used for recognizing the meaning of new, previously unseen visual areas in new documents. Unlike the remaining tagging methods, the visual classification approach requires a training set of documents for setting up the classifier as we show on a practical example in Sect. 9.4. However, the trained classifier may be later used for a whole set of documents coming even from different web sources.

For each tag, there is a single tagger defined that takes into account different criteria. The tagger may combine multiple methods with different supports. For example, the personal names may be recognized by DBPedia concept annotation (with the highest support) but the NER classifier may be used as a fallback solution (with a lower support) for recognizing the names that have no corresponding DBPedia resource.

7   Tag Disambiguation

After the visual areas have been approximately tagged, we disambiguate the tags by considering combinations of the data fields that are expected in the extracted data records (for example considering the *title – authors* or *title – pages* combination in our example in Fig. 1). We assume that all the data records are presented in a visually consistent way in the source document. Based on this assumption, we define presentation constraints on the data records that must apply for considering the records to be visually consistent. Then, the disambiguation task consists of finding the best matching record presentation and layout that meets the visual consistency constraints on one side and covers as many tagged visual areas as possible on the other side.

7.1 Visual Presentation Constraints

For considering the data records to be visually consistent, we require both the consistent presentation style of the individual data fields and consistent layout of the individual fields that form a single data record.

Text Style Consistency. For the individual text fields, we require that the visual areas with the same tag assigned (for example all the paper titles) have the same visual style in the document. We have defined the area style as a tuple of visual features (2). Let's consider a set of set of visual areas A_t that have the tag t assigned and let S_t be a set of styles of all the visual areas in A_t. Then, let n_f be the number of visual features that have equal values for all the styles $s \in S_t$. We say that A_t has a consistent style if n_f is over certain threshold.

Based on our practical experiments, we allow one visual feature that is often used by the document authors to further distinguish the individual records (for example some papers considered to be more important have a bold title or use a different color). Therefore, we use $n_f = 4$ for our experiments.

Content Layout Consistency. The layout consistency constraint is based on our assumption that the layout relationships between the individual data fields expressed by their mutual positions in the page are the same for all data records. For this purpose, we define four relations $R_{side}, R_{after}, R_{below}, R_{under} \subseteq A \times A$ that are defined based on the positions of the areas in the page. Considering a pair of visual areas $a_1, a_2 \in A$ and their respective positions $rect_1, rect_2$ in the page, we define the relations as follows:

- $(a_1, a_2) \in R_{side}$ when a_1 and a_2 are on the same line (their y coordinates overlap), a_2 is placed to the right of a_1 without any other visual area being placed between a_1 and a_2 and the horizontal distance between a_1 and a_2 is not larger than $1\,\text{em}$[1] (shortly, a_2 placed next to a_1).
- $(a_1, a_2) \in R_{after}$ when a_1 and a_2 are on the same line and a_2 is placed to the right of a_1 anywhere on the line (a_2 is on the same line after a_1).
- $(a_1, a_2) \in R_{under}$ when a_1 and a_2 are placed roughly in the same column (their x coordinates overlap) and a_2 is placed below a_1 without any other visual area being placed between a_1 and a_2 and the vertical distance between a_1 and a_2 is not larger than $0.8\,\text{em}$ (a_2 is placed just below a_1).
- $(a_1, a_2) \in R_{below}$ when a_1 and a_2 are placed roughly in the same column (their x coordinates overlap) and a_2 is placed anywhere below a_1.

As we may notice, $R_{side} \subseteq R_{after}$ and $R_{under} \subseteq R_{below}$. For each pair of data fields, we choose the most supported one by trying to cover as many tagged visual areas as possible using each relation. Since one-to-many relationships are allowed between the data fields, any of the above relations may turn out to be the most supported one.

---

[1] In typography, $1\,\text{em}$ is a length corresponding to the point size of the current font.

7.2   Matching the Visual and Semantic Relationships

The tag disambiguation in our approach is based on discovering the most supported combinations of the tagged areas in the page. Considering the target domain described by an ontology (such as our example in Fig. 1), we find the binary relationships with the one-to-many or one-to-one cardinality between the different data properties in the ontology. We assume that the same semantic relationships between two data properties are represented by the same layout relation between the corresponding visual areas for all the data records in the page and in the same time, the visual areas corresponding to the same data type properties have the consistent visual style as defined in Sect. 7.1.

In our sample ontology, we may identify the following one-to-many (or one-to-one) relationships that are expected to have a corresponding visual representation in the document: *section – title, title – author, title – pages*. Note that the paper title may be viewed as a record-identifying field here as defined in [6].

Let's consider a single relationship between the properties represented by the tags t_1 and $t_2 \in T$. Let s_{min} be a minimal value of the tag support (5) that is required for considering the area to have the given tag assigned and let A_{t_1} and A_{t_2} be the sets of visual areas that have the respective tags assigned:

$$A_{t_1} = \{a \in A : ((a, t_1), s) \in tagging \wedge s \geq s_{min}\} \qquad (6)$$
$$A_{t_2} = \{a \in A : ((a, t_2), s) \in tagging \wedge s \geq s_{min}\} \qquad (7)$$

and let S_{t_1} and S_{t_2} be the sets of all styles (2) of the visual areas that belong to A_{t_1} and A_{t_2} respectively. We define a *configuration* of a record extractor as follows:

$$c = (s_{t_1}, s_{t_2}, R) \qquad (8)$$

where $s_{t_1} \in S_{t_1}$, $s_{t_2} \in S_{t_2}$ and R is a layout relation as defined in Sect. 7.1. For each such configuration, we may find a set of matching pairs of visual areas:

$$M_c = \{(a_1, a_2) : a_1 \in A_{t_1} \wedge a_2 \in A_{t_2} \wedge style(a_1) = s_{t_1} \wedge style(a_2) = s_{t_2}$$
$$\wedge (a_1, a_2) \in R\} \quad (9)$$

where $style(a_1)$ and $style(a_2)$ are the styles of a_1 and a_2 respectively. The goal of the tag disambiguation to find a configuration c with the largest set M_c of the corresponding area pairs.

The whole tag disambiguation algorithm for a pair of tags t_1, t_2 corresponding to a one-to-many relationship in the domain ontology may be summarized in the following steps:

1. Compute A_{t_1}, A_{t_2} and the corresponding sets of styles S_{t_1} and S_{t_2} with the minimal support s_{min} set to a higher value (we use $s_{min} = 0.6$ for considering only the tags assigned with some safe support).
2. Compute M_c for all possible configurations c and find the resulting configuration $c_x = (s_{t_1 x}, s_{t_2 x}, R_x)$ where M_{c_x} is the largest set of matching pairs.

3. Decrease s_{min} and recompute A_{t_1}, A_{t_2} and S_{t_1} and S_{t_2} in order to consider even the areas with the tags assigned with a low support (we use $s_{min} = 0.1$).
4. Recompute M_{c_x} for the previously discovered configuration c_x.

After the last step, M_{c_x} contains visually consistent pairs of visual areas (a_1, a_2) that correspond to the same pairs of data fields in the data records.

This process may be generalized to consider multiple one-to-many relationships: we just search for multiple configurations c while maintaining the consistency of s_{t_1} and s_{t_2} and we obtain one set M_{c_x} for each one-to-many relationship. For the one-to-one relationships, the process is equal; the only difference is in the M_c size computation where we consider all the (a_1, a_i) pairs (for all available values of i) as a single pair when computing the size of M_c.

8   Record Extraction

The obtained sets of matches M_c identify the visual areas that contain the corresponding data fields from all the data records discovered in the document. Since the visual areas are directly linked to text boxes from the source document (1), the text content contained in the area may be obtained by a simple concatenation of the text contents of the text boxes.

Depending on the target domain and the area granularity chosen in the preprocessing step (see Sect. 5.1), it may be necessary to further postprocess the extracted text. The postprocessing includes converting the text content to particular data types (such as numbers or dates) or cleaning the text from an additional content. Finally, the obtained values may be mapped to the appropriate ontological properties.

9   Experimental Evaluation

We have implemented the proposed method of data records extraction in Java using our FITLayout framework[2]. The framework is able to process the HTML and PDF input documents by using the CSSBox rendering engine[3]. In order to demonstrate the applicability of the method, we have chosen four application domains, each having some specific features. Although it is not our primary aim to outperform the existing methods in terms of precision, we provide the evaluation of the achieved precision and recall for each sample application in order to show that the obtained results are usable in practice.

9.1   Conference Papers

For the conference paper domain, we have used a dataset from the Semantic Publishing Challenge at the ESWC 2015 conference [5]. The input dataset consists

---

[2] http://www.fit.vutbr.cz/~burgetr/FITLayout/.
[3] http://cssbox.sourcegorge.net/.

Table 1. Results for the conference papers task: number of records extracted, precision, recall and F-measure for two different data sets.

Data set	#rec	P	R	F
(A) Complete dataset (115 documents)	2420	0.976	0.955	0.966
(B) Only documents containing page numbers	883	0.997	0.975	0.986

of 148 selected CEUR workshop proceedings pages[4] from the years 1994–2014 containing the metadata about 2,500+ papers. The input HTML documents are very variable regarding both the code and the visual style. On this dataset, we would like to demonstrate that our approach is able to automatically adapt to the presentation style used in each document and based on the specified domain knowledge, it is able to extract the paper information from a large set of diverse documents.

The extraction task is defined by ontology in Fig. 1 and a set of taggers for assigning the *title*, *authors*, *section* and *pages* tags. For tagging the possible *authors*, we have used the Stanford NER classifier [7] for personal name recognition. The remaining taggers are implemented using regular expressions defining the allowed format of the corresponding data fields.

Since not all the documents contain the page numbers and sections, we have run two experiments: (A) on the complete data set (148 documents) with matching only the *title – authors* pairs and (B) on a subset of documents containing the sections and page numbers (67 documents) with matching the complete records. We have used the evaluation data provided by the SemPub Challenge organizers to evaluate our results and we provide the obtained results in Table 1. As we may notice, we have obtained better results for the (B) dataset which has two main reasons: first, the (B) dataset contains newer documents that are more visually consistent and second, by adding *pages* and *section* tags, the disambiguation is more efficient (more inconsistent combinations may be excluded from the result).

9.2   Sports Results

For the demonstration of the DBPedia concept matching usage, we have chosen the sports results domain as an example of integrating a rapidly changing external data source with DBPedia. We have extracted the records containing athlete *name*, *country* and current *points* from the current tennis and cycling rankings available on the web.

We have used DBPedia Spotlight web service for recognizing the athlete names (the matched DBPedia resource should be instance of `dbo:Athlete`) and countries (instance of `dbo:Country`). Moreover, we have used Stanford NER classifier for recognizing the personal names a locations in case no corresponding resource is available in DBPedia. All visual areas containing a numeric value are considered a possible *points* value and tagged with the corresponding tag.

---

[4] http://ceur-ws.org/.

For every source document[5], we have prepared the "golden standard" data for evaluation by manually transforming the source HTML code to a structured CSV table using a text editor. The results in Table 2 show that based on the assigned tags, our method is able to automatically infer the presentation pattern used for presenting the data records and extract the records with a high precision. In a few cases, the personal names are not identified correctly (there is no corresponding DBPedia resource and the NER classifier failed to recognize the name) which is the reason of lower recall. The resulting extracted records are linked to the corresponding athlete resources in DBPedia. This demonstrates the possibility of an easy integration of an external resource with DBPedia without any predefined templates.

9.3 Timetables

Timetables provide a data source containing an extremely low amount of labels and other additional information that could be used for the data interpretation. Actually, a timetable often contains only the data (hours, minutes, station names) formatted in a specific way leaving its interpretation to a great extent on the experience of the human reader. Motivated by a practical need, we have used the timetables available at the official Czech public transportation timetable portal.[6] The timetables are published here in PDF files (see Fig. 3 for an example) providing a good example of processing data-rich PDF documents.

Table 2. The sports results tasks

Source	#rec	P	R	F
ATP rankings (tennis.com)	200	1.000	0.935	0.966
WTA rankings (tennis.com)	200	1.000	0.925	0.961
Road cycling rankings (uci.ch)	2488	1.000	0.933	0.965
Mountain bike rankings (uci.ch)	1627	1.000	0.978	0.989

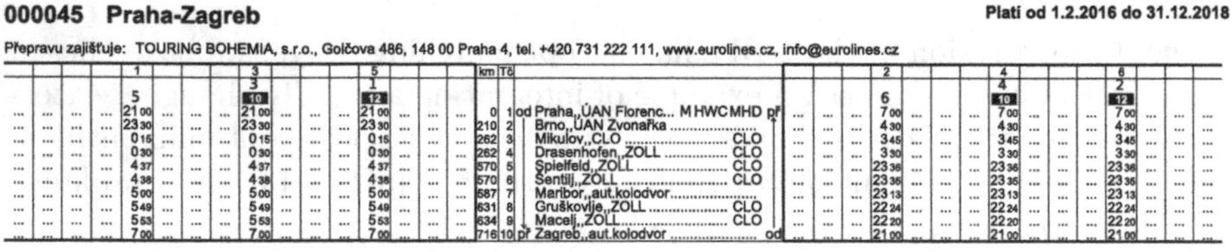

Fig. 3. A sample timetable

[5] The URLs of the source documents were http://www.tennis.com/rankings/ATP/, http://www.tennis.com/rankings/ATP/, http://www.uci.ch/road/ranking/ and http://www.uci.ch/mountain-bike/ranking/ respectively.

[6] http://portal.idos.cz.

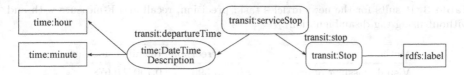

Fig. 4. Ontology used for timetables. The concepts and properties come from the OWL Time ontology and Transit ontology

The domain knowledge is represented by an ontological description in Fig. 4 and taggers for the tags *hours* and *minutes* based on the recognition of numbers in the corresponding range and for *stops* (stop names) based on matching with a fixed list of existing stops (which is available in this domain) combined with regular expressions used when the matching fails.

We have tested our method on 30 different time table documents from the above mentioned portal. The extracted data was compared with a golden standard that was created manually by transforming the PDF files to CSV data using a semi-automatic transformation based on regular expressions tailored for the particular documents. Because the time tables contain a large amount of $(hour, minute, stop)$ records (we have obtained 5130 records in total), the tag disambiguation works very efficiently in this case and we have extracted all the records correctly $(P = R = 1.0)$. It is worth noting that all the *hour* and *minute* pairs have been identified correctly although the initial tagging is very ambiguous (many visual areas share both tags after the initial text classification).

9.4   News Articles

We have chosen the news articles domain to demonstrate a different application scenario. Unlike the documents in the previous domains that typically contained many data records (papers or times), the news web pages usually contain one full article in a document. However, each news website contains many such documents that follow a visually consistent presentation style. Therefore, we may treat a set of documents as a single input page containing multiple articles.

For this task, we view the individual news articles as data records containing data fields that we have assigned the following tags: *title* (article title), *author* (author name), *pubdate* (date of publication) and *paragraph* (a paragraph of text). Considering the title to be the record-identifying field, the *title – paragraph* pairs correspond to a one-to-many relation, the *title – author* and *title – pubdate* pairs are one-to-one relations.

Due to the specific properties of the news domain where it may be difficult to recognize the individual parts such as titles and subtitles by text classification only, we employ a visual classification approach that allows to assign the tags to the areas based on their visual appearance. This approach that we have presented in detail in [3] uses the visual features of the individual visual areas: Font features such as the font size, weight and style, spatial features (position in the page and size), text features (numbers of characters and lines) and color features

Table 3. Results for the news articles task: precision, recall and F-measure with and without using tag disambiguation

Method	Precision	Recall	F-measure
Visual classifier only	0.593	0.790	0.678
Visual classifier + disambiguation	0.978	0.986	0.982

(luminosity, contrast). The values of the features are expressed numerically and used as an input for a generic classifier[7]. Therefore, in contrast to the other applications presented in the previous sections, a training set of documents is required for setting up the classifier. Later in the classification step, the trained classifier directly assigns tags to the visual areas in new documents.

For testing, we have used the news articles on reuters.com and cnn.com news portals. We have taken 30 documents with articles from each website. We have manually annotated the source documents by manually assigning the appropriate tags to the individual visual areas in the documents.[8] Then, 5 documents from each web site were used for training the classifiers (one for each source website) based on the visual features of the manually tagged areas. Later, the trained classifiers were used for assigning tags to all the visual areas in the complete dataset from the given website.

The results obtained are shown in Table 3. The first row shows the values obtained by comparing the classification results with the manually assigned tags. This corresponds to the scenario presented in [3]. The second line shows the result with disambiguation where the visual classification results were used as the initial tagging for the tag disambiguation process described in Sect. 7. As we may see, the disambiguation greatly improves the resulting precision and recall.

10   Conclusions

We have presented a record extraction approach from web documents that is based on searching the most frequent visual presentation patterns in the documents while assuming that multiple instances of the records are available in the documents. The extraction itself is based only on the knowledge available for the target domain that includes the expected structure of the extracted records and an estimation of possible values (or alternatively a style) of the data fields. We consider this as the main benefit of the presented approach. As the result, the method is independent on the source document format, and it does not rely on any kind of templates used or labels or annotations present in the source documents. The experimental results demonstrate the applicability of the approach for different scenarios and document sources.

---

[7] For our experiments, we have used the J.48 classifier from the WEKA package (which is an implementation of the C4.5 decision tree classifier) mainly for its speed.

[8] The used FITLayout framework provides a graphical annotation tool that was used for this task.

Acknowledgments. This work was supported by The Ministry of Education, Youth and Sports from the National Programme of Sustainability (NPU II); project IT4Innovations excellence in science – LQ1602.

References

1. Anderson, N., Hong, J.: Visually extracting data records from query result pages. In: Ishikawa, Y., Li, J., Wang, W., Zhang, R., Zhang, W. (eds.) APWeb 2013. LNCS, vol. 7808, pp. 392–403. Springer, Heidelberg (2013). doi:10.1007/978-3-642-37401-2_40
2. Burget, R.: Hierarchies in HTML documents: linking text to concepts. In: 15th International Workshop on Database and Expert Systems Applications, pp. 186–190. IEEE Computer Society (2004)
3. Burget, R., Burgetová, I.: Automatic annotation of online articles based on visual feature classification. Int. J. Intell. Inf. Database Syst. **5**(4), 338–360 (2011)
4. Daiber, J., Jakob, M., Hokamp, C., Mendes, P.N.: Improving efficiency and accuracy in multilingual entity extraction. In: Proceedings of the 9th International Conference on Semantic Systems (I-Semantics) (2013)
5. Iorio, A.D., Lange, C., Dimou, A., Vahdati, S.: Semantic publishing challenge – assessing the quality of scientific output by information extraction and interlinking. In: Gandon, F., Cabrio, E., Stankovic, M., Zimmermann, A. (eds.) SemWebEval 2015. CCIS, vol. 548, pp. 65–80. Springer, Cham (2015). doi:10.1007/978-3-319-25518-7_6
6. Embley, D.W., Campbell, D.M., Jiang, Y.S., Liddle, S.W., Lonsdale, D.W., Ng, Y.K., Smith, R.D.: Conceptual-model-based data extraction from multiple-record web pages. Data Knowl. Eng. **31**(3), 227–251 (1999)
7. Finkel, J.R., Grenager, T., Manning, C.: Incorporating non-local information into information extraction systems by Gibbs sampling. In: Proceedings of the 43rd Annual Meeting on Association for Computational Linguistics, ACL 2005, pp. 363–370 (2005)
8. Goh, P.L., Hong, J.L., Tan, E.X., Goh, W.W.: Region based data extraction. In: 2012 9th International Conference on Fuzzy Systems and Knowledge Discovery (FSKD), pp. 1196–1200, May 2012
9. Hong, J.L., Siew, E.G., Egerton, S.: Information extraction for search engines using fast heuristic techniques. Data Knowl. Eng. **69**(2), 169–196 (2010). doi:10.1016/j.datak.2009.10.002
10. Kolchin, M., Kozlov, F.: A template-based information extraction from web sites with unstable markup. In: Presutti, V., Stankovic, M., Cambria, E., Cantador, I., Di Iorio, A., Di Noia, T., Lange, C., Reforgiato Recupero, D., Tordai, A. (eds.) SemWebEval 2014. CCIS, vol. 475, pp. 89–94. Springer, Cham (2014). doi:10.1007/978-3-319-12024-9_11
11. Milicka, M., Burget, R.: Information extraction from web sources based on multi-aspect content analysis. In: Gandon, F., Cabrio, E., Stankovic, M., Zimmermann, A. (eds.) SemWebEval 2015. CCIS, vol. 548, pp. 81–92. Springer, Cham (2015). doi:10.1007/978-3-319-25518-7_7
12. Su, W., Wang, J., Lochovsky, F.H.: ODE: ontology-assisted data extraction. ACM Trans. Database Syst. **34**(2), 121–1235 (2009). doi:10.1145/1538909.1538914

13. Weng, D., Hong, J., Bell, D.A.: Extracting data records from query result pages based on visual features. In: Fernandes, A.A.A., Gray, A.J.G., Belhajjame, K. (eds.) BNCOD 2011. LNCS, vol. 7051, pp. 140–153. Springer, Heidelberg (2011). doi:10.1007/978-3-642-24577-0_16

14. Weng, D., Hong, J., Bell, D.A.: Automatically annotating structured web data using a SVM-based multiclass classifier. In: Benatallah, B., Bestavros, A., Manolopoulos, Y., Vakali, A., Zhang, Y. (eds.) WISE 2014. LNCS, vol. 8786, pp. 115–124. Springer, Cham (2014). doi:10.1007/978-3-319-11749-2_9

15. Zheng, X., Gu, Y., Li, Y.: Data extraction from web pages based on structural-semantic entropy. In: Proceedings of the 21st International Conference on World Wide Web, WWW 2012 Companion, pp. 93–102. ACM, New York (2012). doi:10.1145/2187980.2187991

Statistical Induction of Coupled Domain/Range Restrictions from RDF Knowledge Bases

Basil Ell[✉], Sherzod Hakimov, and Philipp Cimiano

CIT-EC, Universität Bielefeld, Bielefeld, Germany
{bell,shakimov,cimiano}@cit-ec.uni-bielefeld.de

Abstract. Statistical Schema Induction can be applied on an RDF dataset to induce domain and range restrictions. We extend an existing approach that derives independent domain and range restrictions to derive coupled domain/range restrictions, which may be beneficial in the context of Natural Language Processing tasks such as Semantic Parsing and Entity Classification. We provide results from an experiment on the DBpedia graph. An evaluation shows that high precision can be achieved. Code and data are available at https://github.com/ag-sc/SchemaInduction.

Keywords: Statistical Schema Induction · RDF · Property restrictions

1 Introduction

RDFS domain restrictions and range restrictions for a property let us infer to which class the subject and the object, respectively, of a triple with this property belong to. For example, given the domain restriction (foaf:knows, rdfs:domain, foaf:Person) and (foaf:knows, rdfs:range, foaf:Person), from the triple (ex:Frank, foaf:knows, ex:Vidya) we can infer the class assertions (ex:Frank, rdf:type, foaf:Person) and (ex:Vidya, rdf:type, foaf:Person).

In the context of the Semantic Web knowledge representation format RDF, *Statistical Schema Induction* is the process of inducing ontological statements such as RDFS statements and OWL statements from RDF data, such as domain and range restrictions or subclass relations.

Domain restrictions and range restrictions were so far considered independently [9]. For example, RDFS does not allow to specify that given a statement (ex:s, ex:p, ex:o), if ex:s belongs to class c_1, then ex:o belongs to class c_2. This makes sense since property restrictions are entailment rules and not constraints. However, when applying them as heuristics instead, coupling domain and range restrictions becomes interesting. For example, consider the DBpedia property dbo:champion with domain dbo:SportsEvent and range dbo:Athlete. When observing concrete data, one can see that this property is either used for subjects of class dbo:SportsEvent and objects of class dbo:Athlete or, among other cases, for subjects of class dbo:GolfTournament and objects of class dbo:GolfPlayer. Having identified a golf tournament in text near and entity identified as a person, one may now guess that the person is a golf player.

© Springer International Publishing AG 2017
M. van Erp et al. (Eds.): ISWC 2016 Workshops, LNCS 10579, pp. 27–40, 2017.
https://doi.org/10.1007/978-3-319-68723-0_3

In this paper we propose an approach based on Frequent Itemset Mining (FIM) to statistically induce coupled domain and range restrictions. Given a knowledge base encoded in RDF, for a property p our method creates a set of statements of the form (D, R, c) where D is the set of domain classes, R is the set of range classes, and $c \in [0, 1]$ is a support value.

We envision that the application of independent and coupled domain/range restrictions is interesting in several scenarios. We present details on Semantic Parsing and Entity Classification below.

1. Semantic Parsing: Question answering may consist of the task of mapping natural language questions to SPARQL queries that can then be evaluated over an RDF dataset. An example of a question and the corresponding query from QALD-6[1] is given below: "Which actors were born in Germany?"

```
PREFIX rdf: <http://www.w3.org/1999/02/22-rdf-syntax-ns#>
PREFIX dbo: <http://dbpedia.org/ontology/>
PREFIX dbr: <http://dbpedia.org/resource/>
SELECT ?uri WHERE {
    ?uri rdf:type dbo:Actor .
    ?uri dbo:birthPlace dbr:Germany .
}
```

Listing 1. SPARQL query from QALD-6

The string "born" can be interpreted as mapping to a property. In DBpedia there are various candidates for the interpretation of "born": dbo:birthDate, dbo:birthYear, and dbo:birthPlace. For the question above, only the interpretation in terms of dbo:birthPlace makes sense as Germany matches to the range restriction of dbo:birthPlace because the range of the property dbo:birthPlace is dbo:Place and dbr:Germany is an instance of the class dbo:Place. As the entity dbr:Germany does not comply with the range restrictions of dbo:birthYear nor of dbo:birthDate, these interpretations can be ruled out by a Question Answering system that takes into account domain and range restrictions. Using coupled domain/range restrictions can help to rule out further interpretations in terms of properties and entities.

2. Entity Classification: given are a property with two different pairs of domain/range classes: (c_1, c_2) and (c_3, c_4). That means, c_1 and c_2 can occur together as domain and range, respectively, and c_3 and c_4 can occur together as domain and range, respectively. If we find an entity, e.g. in text, that is an instance of the class c_1, then, if the object of the relation is also found in text but is ambiguous since there are multiple entities that match, then the entity that is an instance of the class c_2 is more likely to be the correct one than entities of other types, e.g., of those that are instances of the class c_4. This is an example for the application of coupled domain/range restrictions.

---

[1] See http://qald.sebastianwalter.org/index.php?x=benchmark&q=6.

The main contributions of this paper are:

- We motivate the idea of coupled domain/range restrictions that can serve as heuristics in NLP tasks such as Question Answering and Entity Classification.
- We formalize an existing approach for the induction of independent domain and range restrictions as well as our approach for the induction of coupled domain/range restrictions.
- We carry out an experiment on DBpedia and make all induced (independent and coupled) restrictions available to the community.

The remainder of this paper is structured as follows. We give a quick introduction to Frequent Itemset Mining as well as relevant vocabulary (RDF, RDFS, and OWL) in Sect. 2, present the existing method to derive domain and range restrictions independently as introduced by [9] as well as our method in Sect. 3, describe an experiment to induce coupled domain range restrictions from DBpedia in Sect. 4, evaluate the outcome of the experiment in Sect. 5, discuss related work in Sect. 6, and, finally, conclude in Sect. 7.

2 Preliminaries

In this section we briefly introduce the graph-based data model RDF, relevant terms from the RDF(S) vocabularies[2] and from the OWL[3] vocabulary, as well as Frequent Itemset Mining.

2.1 RDF, RDFS, and OWL

An RDF graph consists of a set of *RDF triples*. An RDF triple $t = (s, p, o) \in (\mathcal{U} \cup \mathcal{B}) \times (\mathcal{U}) \times (\mathcal{U} \cup \mathcal{B} \cup \mathcal{L})$ is an ordered set consisting of a subject s, a predicate p, and an object o. $\mathcal{U}$ is a set of URIs, $\mathcal{B}$ is a set of blank nodes (existentially qualified variables), and L is a set of literals. $\mathcal{U}$, $\mathcal{B}$, and $\mathcal{L}$ are pairwise disjunct.

From the RDF, RDFS, and OWL vocabularies a small set of terms is relevant in the context of the current work: `rdf:type`, `rdfs:domain`, `rdfs:range`, `rdfs:subClassOf`, and `owl:equivalentClass`.

The property `rdf:type` is used for explicit class assertions. For example, the RDF triple (`ex:X, rdf:type, ex:C`) explicitly expresses that `ex:X` is a member of the class `ex:C`. With the triples (`ex:P, rdfs:domain, ex:C`) and (`ex:P, rdfs:range, ex:D`) a domain restriction (first triple) and a range restriction (second triple) are specified. Given these domain and range restrictions, from a statement, such as (`ex:A, ex:P, ex:B`), the two class assertions (`ex:A, rdf:type, ex:C`) and (`ex:B, rdf:type, ex:D`) can be derived (by taking into account RDFS semantics[4]) – the first triple via the domain restriction and the second triple via the range restriction. Subclass relations between classes can

---

[2] See http://www.w3.org/TR/rdf-primer/.
[3] See http://www.w3.org/TR/owl-features/.
[4] See http://www.w3.org/TR/2004/REC-rdf-mt-20040210/.

be expressed using the property `rdfs:subClassOf`. For example, from a statement (`ex:C`, `rdfs:subClassOf`, `ex:D`) it follows that `ex:C` is a subclass of `ex:D`, which means that every entity that is a member of class `ex:C` is also a member of class `ex:D`. Equivalence between two classes can be expressed via the symmetric property `owl:equivalentClass`, as in (`ex:C`, `owl:equivalentClass`, `ex:D`).

2.2   Frequent Itemset Mining

Let I be a set of items. Given a list T of subsets of the item set (also known as a set of transactions) and given a value $\tau \in [0, 1]$ (also known as the support threshold), the objective of frequent itemset mining consists of creating a set of itemsets O where each set $o \in O$ has a support of greater than or equal to τ, which means that it is subset to at least $\tau * |T|$ itemsets in T. Support of an itemset s is calculated as the number of itemsets $t \in T$ where $s \subseteq t$ devided by $|T|$. See [7] for a textbook-style introduction to Frequent Itemset Mining.

As a classical example, let I be the set of articles available in a grocery store and let T be the set of sets of articles in individual shopping baskets. Frequent Itemset Mining [2] applied on T then tells us which articles are frequently (depending on the τ value) purchased together. Given $\tau = 0.9$, each set of articles can be found in at least 90% of the shopping baskets.

Note that for each frequent itemset, all of its subsets are also frequent itemsets. This property is referred to as the monotonicity of frequent itemsets. *Frequent Maximal Itemset Mining* is the task of deriving only those frequent itemsets that are maximal, which means that a frequent maximal itemset is not a true subset of another frequent itemset.

3   Method

In this section we describe two methods. The first method is the state-of-the-art method by Völker and Niepert [9] that derives independent domain and range statements from an RDF graph using the Frequent Itemset Mining tool [6]. The second method, which is a new contribution, derives coupled domain/range statements from an RDF graph using Frequent Itemset Mining.

Note that we do not intend to provide a method that outperforms the one presented in [9] but rather introduce the new problem of inducing coupled domain/range restrictions for which we introduce a method that builds on the method by Völker and Niepert.

Input to both methods is an RDF graph G, a set of properties P, and a support threshold value τ.

Output for the first method is a set of (`domain classes`, `support`) tuples and a set of set of (`range classes`, `support`) tuples. Output for the second method is a set of (`domain classes`, `range classes`, `support`) tuples.

The examples that we present to illustrate the method were created by querying against the DBpedia [1] dataset.[5] We ignore classes that are not in the DBpedia namespace, such as `owl:Thing`.

3.1 Deriving Independent Domain and Range Statements

Here we describe the core of the method based on Frequent Itemset Mining to induce domain and range axioms as introduced by Völker and Niepert [9]. The method to gather data from an endpoint is a bit different, as we will explain later in Sect. 6.

We create a class dictionary D which is an injection that assigns an integer value to each class in G. Given an RDF graph G, the class dictionary D, and a set of properties P, for each property $p \in P$ we create two transaction tables T_d^p and T_r^p as follows. For each triple $(s, p, o) \in G$ we add a row to T_d^p containing all members of c_s, which is the set of classes the entity s belongs to. We do not add a row to T_d^p if c_s is empty. Furthermore, we add a row to T_r^p containing all members of c_o, which is the set of class identifiers of the classes the object entity o belongs to or the set of datatypes the object literal o belongs to. We do not add a row to T_r^p if c_o is empty. The sets of classes do not only contain the explicit classes c of an entity e as defined via the triple (`e, rdf:type, c`), but also all superclasses of c via the transitive relation *rdfs:subClassOf*, all their equivalent classes via the transitive relation *owl:equivalentClass*, as well as all implicit types that can be inferred via existing domain or, respectively, range restrictions.

For example, given the property `dbo:author` and the triple (`dbo: Gantenbein, dbo:author, dbr:Max_Frisch`), for the resource `dbo:Gantenbein` we obtain the set $c_s = $ {`dbo:Book, dbo:Work, dbo:WrittenWork`}. All three classes happen to be available via direct class assertions. `dbo:Work` is the domain of the property (given in the DBpedia Ontology). The superclasses of the directly asserted classes are `dbo:Book`, `dbo:Work`, and `dbo:WrittenWork` and the superclass of the domain class is `dbo:Work`.

For the resource `dbr:Max_Frisch` we obtain the set $c_o = $ {`dbr:Writer, dbo:Person, dbo:Agent`}. Again, all three classes happen to be available via direct class assertions. `dbo:Person` is the range of the property. The superclasses of the directly asserted classes are `dbo:Writer`, `dbo:Agent`, and `dbo:Person` and the superclasses of the range class are `dbo:Person` and `dbo:Agent`. To the domain transaction table $T_d^{dbo:author}$ we would therefore add a line such as "0 1 2", given that the class identifiers refer to the classes `dbo:Book`, `dbo:Work`, and `dbo:WrittenWork`, respectively, in the class dictionary D. This line in $T_d^{dbo:author}$ would express that there is a triple with the property `dbo:author` where the subject belongs to the classes with the identifiers 0, 1, and 2.

---

[5] The prefixes `dbo` and `dbr` refer to http://dbpedia.org/ontology/ and http://dbpedia.org/resource/, respectively.

Given a transaction table and a support threshold τ, we perform frequent maximal[6] itemset mining to derive a set of classes and their support values where the support values are not less than τ. We reduce each set of classes so that for each class all of its (implicit and explicit) superclasses are removed from the set. Moreover, if a set contains multiple equivalent classes, then all but the first in lexicographic order are removed. For example, the set {dbo:Athlete, dbo:Person, dbo:Agent} is reduced to {dbo:Athlete}, since this class is a subclass of the two other classes. The purpose of adding superclasses and equivalent classes in the first place is, that within a knowledge graph such as DBpedia, not all entities are necessarily consistently typed. For example, sometimes a superclass is explicitly given, sometimes it is not. Adding them leads to more consistent entries in the transaction table.

For example, given a domain transaction table $T_d^{dbo:author}$ that was created with 10,000 triples and $\tau = 0.5$ we obtained three frequent maximal itemsets:

- ({dbo:Work}, 10000/10000)
- ({dbo:WrittenWork}, 7771/10000)
- ({dbo:Book}, 6396/10000)

From these we can create three domain restrictions:

- (dbo:author, rdfs:domain, dbo:Work)
- (dbo:author, rdfs:domain, dbo:WrittenWork)
- (dbo:author, rdfs:domain, dbo:Book)

Note that this set of domain restrictions contains redundancies. Given the third restriction, the first two restrictions could automatically be created since dbo:Work and dbo:WrittenWork are superclasses of dbo:Book. Therefore, we reduce the output to the restriction with the most specific class. Thus, the restriction (dbo:author, rdfs:domain, dbo:Book) is the only itemset remaining after reduction.

3.2   Deriving Coupled Domain and Range Statements

We create a class dictionary D which is an injection as follows. For each class c in G we create two new strings "domain=" + c and "range" + c via string concatenation and assign different integer values in the dictionary. Given an RDF graph G, the class dictionary D, and a set of properties P, for each property $p \in P$ we create one transaction table T_{dr}^p as follows. For each triple $(s, p, o) \in G$ we add a row (transaction) to T_{dr}^p containing all members of $c_s' \cup c_o'$ As for the other method above, c_s is the set of classes the entity s belongs to and c_o is the set of classes the object entity o belongs to or the set of datatypes the object literal o belongs to and the sets c_s' and c_o' are derived from c_s and c_o, respectively, as follows. For each member $c \in c_s$ (c_o, respectively), we concatenate the string

---

[6] Note that the authors of [9] do not explicitly mention that they derive frequent *maximal* itemsets only. But since non-maximal itemsets, such as empty itemsets, are irrelevant, we assume they perform frequent maximal itemset mining.

"domain=" ("range=", respectively) and c and add the resulting string to c'_s (c'_o, respectively). The sets of classes do not only contain the explicit classes c of an entity e as defined via the triple (e, rdf:type, c), but also all superclasses of c via the transitive relation *rdfs:subClassOf*, all their equivalent classes via the transitive relation *owl:equivalentClass*, as well as all implicit types that can be inferred via existing domain or, respectively, range restrictions.

From a transaction table T^p_{dr} we derive frequent maximal itemsets and reduce them as carried out in the method above. Depending on whether the class names begin with the string "domain=" or "range=" we can distribute them to the set of domain classes and the set of range classes.

An example of a frequent itemset is (dbo:bronzeMedalist, {dbo:Olympic Event}, {dbo:Person}, 5200/10000). From this itemset we can create two restrictions: (dbo:bronzeMedalist, rdfs:domain, dbo:OlympicEvent) and (dbo:bronzeMedalist, rdfs:range, dbo:Person). However, if these restrictions are represented in this form and are added to an RDF graph, then the domain restrictions and the range restrictions are not coupled anymore.

4   Experiment

For our experiment we set up a SPARQL endpoint containing DBpedia (version 2015-10). The repository contains 8.8 billion triples, 739 classes, 1099 object properties, and 1734 datatype properties.

For some properties DBpedia already contains domain and range restrictions, as listed in Table 1. The headers of the table denote whether properties have domain or range restrictions, e.g., as in the example $D \wedge \neg R$, that the property has a domain restriction but no range restriction.

We induced restrictions for 1099 object properties and for 1734 datatype properties (see Table 1) for values of τ in the range {1.0, 0.9, 0.8, 0.7, 0.6, 0.5, 0.4, 0.3, 0.2, 0.1} independently of whether domain or range restrictions already exist in DBpedia. Figure 1 shows: (i) the number of domain and range itemsets that were induced for different values of τ (e.g., 2290 domain itemsets (= domain restrictions) were induced for $\tau = 1.0$ and 920 range itemsets (= range restrictions) were induced for $\tau = 0.9$), (ii) the minimum, average, and maximal number of classes in itemsets in domain and range, and (iii) the minimum, average, and maximum number of itemsets induced per property. It is no surprise that the number of domain itemsets and range itemsets grows when τ is decreased. Also, the number of classes within a frequent itemset and the number of frequent itemsets per property grow when τ is decreased.

Table 1. Statistics about Object & Data Properties in DBpedia version 2015-10.

Property type	Total	$D \wedge R$	$\neg D \wedge R$	$D \wedge \neg R$	$\neg D \wedge \neg R$
Object properties	1099	704	120	206	69
Datatype properties	1734	1497	237	0	0

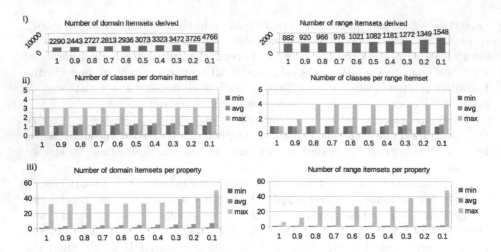

Fig. 1. Basic statistics of induced independent domain and range restrictions.

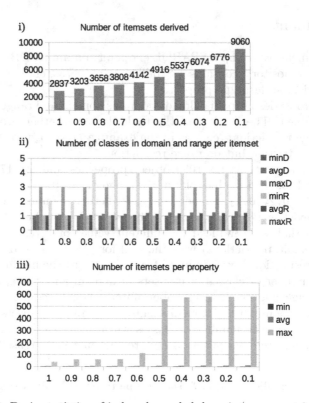

Fig. 2. Basic statistics of induced coupled domain/range restrictions.

Results from inducing coupled domain/range restrictions are shown in Fig. 2. In detail, it shows: (i) the number of itemsets that were induced for different values of τ (e.g.,), (ii) the minimum, average, and maximal number of classes in itemsets in domain and range (e.g., 3203 itemsets (= coupled domain/range restrictions were induced for $\tau = 0.9$), $minD$, stands for the minimum number of classes in the sets of domain classes), and (iii) the minimum, average, and maximum number of itemsets induced per property.

5 Evaluation

We evaluate the results of both methods:

1. We compare the induced domain (range) restrictions against the domain (range) restrictions that already exist in DBpedia, which we treat as gold standard.
2. We compare the induced coupled domain/range restrictions against the independent domain and range restrictions that already exist in DBpedia, which we treat as gold standard.

To compare against the gold standard of DBpedia, we selected 704 object properties and 1497 datatype properties for which both domain restriction and range restriction are known.

5.1 Evaluation of Independent Domain and Range Restrictions

For each domain (range) restriction for which a gold domain (range) restriction exists we compare the induced set of classes against the gold set of classes. This comparison may result in four cases: (i) the sets of classes are identical, (ii) all induced classes are more specific than all gold classes, (iii) all induced classes are less specific than all gold classes, and (iv) else (some classes are more specific, some are less specific, some are neither less nor more specific when compared to the gold set). In Table 2 these cases are denoted with $=$, $>$, $<$, and x, respectively.

We define $precision@\tau$ as the number of frequent maximal itemsets derived with the support threshold τ where either the set of induced classes is the same as the gold classes or where the induced classes are less specific, divided by the number of all frequent maximal itemsets derived with the support threshold τ. The arrows ($\uparrow$) in Table 2 mark the columns where this is the case (i.e., $=$ and $<$).

Table 3 shows an example for each of the four cases (i) the induced domains and the gold domains are identical, (ii) the induced domains are more specific than gold domains, (iii) the induced domains are less specific than the gold domains, and (iv) otherwise, denoted with $=$, $>$, $<$, and x, respectively.

Table 2. Induced independent domain and range restrictions, the frequency of cases, and precision values.

τ	Domain				Range				Domain precision@τ	Range precision@τ
	$=$	$>$	$<$	x	$=$	$>$	$<$	x		
1	722	152	1254	162	723	4	155	0	0.86	1.00
0.9	723	249	1254	217	723	20	155	22	0.81	0.95
0.8	723	268	1254	482	723	30	155	58	0.72	0.91
0.7	723	294	1252	544	723	40	155	58	0.70	0.90
0.6	723	319	1252	642	723	52	155	91	0.67	0.86
0.5	723	349	1252	748	723	57	155	147	0.64	0.81
0.4	723	393	1253	954	723	62	155	241	0.59	0.74
0.3	723	424	1254	1071	723	78	155	316	0.57	0.69
0.2	723	477	1254	1272	723	94	155	377	0.53	0.65
0.1	723	574	1248	2221	722	144	151	531	0.41	0.56
	↑		↑		↑		↑			

Table 3. Examples of induced domain classes and the corresponding gold domain classes. The entries for the column *case* correspond to the comparison of induced domains to gold domains, where four cases are possible: $=$ (the induced domain classes are identical with the gold domain classes), $>$ (the induced domain classes are more specific than gold domain classes), $<$ (the induced domain classes are less specific than gold domain classes), and x (otherwise).

Case	τ	Property	Support	Induced domain	Gold domain
$=$	1.0	dbo:leftTributary	4881/4881	dbo:River	dbo:River
$>$	0.5	dbo:composer	5847/10.000	dbo:TelevisionShow	dbo:Work
$<$	1.0	dbo:chef	54/54	dbo:ArchitecturalStructure	dbo:Restaurant
x	0.8	dbo:launchSite	45/509	dbo:MeanOfTransportation	dbo:SpaceMission

5.2 Evaluation of Coupled Domain/Range Restrictions

Induced coupled domain/range restrictions were evaluated similarly to the independent domain and range restrictions. However, the number of cases that may occur is not 4 ($\{=,>,<,x\}$) but instead 4*4 ($\{=,>,<,x\}\times\{=,>,<,x\}$). Table 4 shows an example for each of the 16 possible cases of itemsets with coupled domain and range.

We define *precision*@τ as the number of frequent maximal itemsets derived with the support threshold τ where either the set of induced domain classes is the same or less specific than the gold domain classes and where the induced range classes is the same or less specific than the gold range classes, divided by the number of all frequent maximal itemsets derived with the support threshold τ. The arrows (↑) in Table 5 mark the columns where this is the case (i.e., $==$, $=<$, $<=$, and $<<$). Table 6 shows the precision@τ values.

Note that the precision of induced coupled restrictions (Table 2) tends to be below the precision of induced independent restrictions (Table 6). The main reason for the lower precision is that domain and range classes of coupled

Table 4. Examples of induced coupled domain and range classes and the corresponding gold classes. The namespace prefix (dbo) has been consistently omitted. d corresponds to domain restriction and r to range restriction. The entries for the column *case* correspond to the comparison of induced domains to gold domains and induced ranges to gold ranges, respectively, where for each comparison four cases are possible: = (the induced classes are identical with the gold classes), > (the induced classes are more specific than gold classes), < (the induced classes are less specific than gold classes), and x (otherwise).

Case	τ	Property	Support	Induced coupled dom./range	Gold domain/range
xx	0.3	launchSite	156/509	d=SocietalEvent, MeanOfTransportation r=MilitaryStructure	d=SpaceMission r=Building
$x<$	0.8	launchSite	455/509	d=SpaceMission, MeanOfTransportation r=ArchitecturalStructure	d=SpaceMission r=Building
$x>$	0.1	writer	1195/10000	d=Wikidata:Q11424,Work r=Writer	d=Work r=Person
$x=$	0.4	militaryBranch	4065/10000	d=Organisation, MilitaryPerson r=MilitaryUnit	d=MilitaryPerson r=MilitaryUnit
$<x$	0.3	militaryBranch	5042/10000	d=Agent r=MilitaryUnit,Place	d=MilitaryPerson r=MilitaryUnit
$<<$	0.7	champion	1349/1349	d=SocietalEvent r=Agent	d=SportsEvent r=Athlete
$<>$	0.2	militaryBranch	10000/10000	d=Person r=Agent	d=MilitaryPerson r=MilitaryUnit
$<=$	0.1	militaryBranch	10000/10000	d=Agent r=MilitaryUnit	d=MilitaryPerson r=MilitaryUnit
$>x$	0.3	dam	198/450	d=RaceHorse r=RaceHorse,Eukaryote	d=Animal r=Animal
$><$	0.2	dam	450/450	d=Mammal r=Species	d=Animal r=Animal
$>>$	0.9	champion	1252/1349	d=Tournament r=GolfPlayer	d=SportsEvent r=Athlete
$>=$	0.8	champion	1348/1349	d=GolfTournament r=Athlete	d=SportsEvent r=Athlete
$=x$	0.3	launchSite	191/509	d=SpaceMission r=MilitaryStructure	d=SpaceMission r=Building
$=<$	0.2	dam	439/450	d=Mammal r=Mammal	d=Animal r=Animal
$=>$	0.1	dam	439/450	d=Animal r=Horse	d=Animal r=Animal
$==$	1.0	launchSite	509/509	d=SpaceMission r=Building	d=SpaceMission r=Building

Table 5. Induced coupled domain/range restrictions and the frequency of cases.

τ	==	=>	=<	= x	>=	>>	><	> x	<=	<>	<<	< x	x =	x >	x <	xx
1.0	721	4	157	0	153	4	32	0	1252	10	326	0	165	0	13	0
0.9	723	20	156	22	249	20	56	16	1256	51	327	66	219	0	18	4
0.8	723	30	157	58	269	30	59	24	1254	75	326	134	481	0	34	4
0.7	723	40	155	58	294	35	67	24	1256	101	326	135	545	1	44	4
0.6	723	52	156	91	319	39	69	25	1257	127	327	207	643	0	99	8
0.5	723	58	156	147	350	42	75	48	1256	140	330	306	749	22	122	392
0.4	723	62	158	242	393	50	78	88	1254	144	326	438	954	23	164	440
0.3	723	78	158	316	425	68	92	91	1256	159	328	520	1071	74	186	529
0.2	723	94	156	378	479	78	105	101	1255	203	326	672	1272	135	196	603
0.1	722	146	152	531	574	105	120	156	1250	314	311	1084	2225	239	348	783
	↑		↑						↑		↑					

Table 6. Precision@τ for induced coupled domain/range restrictions.

τ	1.0	0.9	0.8	0.7	0.6	0.5	0.4	0.3	0.2	0.1
Precision@τ	0.87	0.77	0.67	0.65	0.59	0.50	0.44	0.41	0.36	0.27

domain/range restrictions are often more specific than the gold domain and range classes. See for example Table 4 case >>. Adding restrictions with classes that are too specific to a knowledge graph would result in wrong entailment. For example, by adding the restriction (dbo:champion, rdfs:range, dbo:GolfPlayer), from every triple with that property we can then infer via RDFS entailment that the object of the triple is an instance of the class dbo:GolfPlayer – in other words: every champion is a golf player. However, these specific classes are helpful as heuristics for certain Natural Language Processing tasks as motivated in the introduction.

6   Related Work

Our approach is an extension of a method for statistical schema induction from RDF data introduced by Völker and Niepert [9] which we discuss in detail in Sect. 3.1. Besides inducing domain and range restrictions of properties, in their work further axioms are induced such as subsumption axioms (e.g., (ex:A, rdfs:subClassOf, ex:B)) and transitivity axioms (e.g., (ex:P, rdf:type, owl:TransitiveProperty)). This work was subsequently extended in [4,5] with further types of axioms. The main difference to this work is that we induce independent domain and range restrictions as well as coupled domain/range restrictions whereas in these works only independent domain and range restrictions are induced. However, all methods are based on Frequent Itemset Mining and only differ in some technical details, such as, how the set of classes an entity belongs to are created (e.g., whether equivalent classes (via `owl:equivalentClass`), superclasses (via `rdfs:subClassOf`), and existing domain and range restrictions are regarded). In both approaches the sets of classes are approximated and it cannot be guaranteed that the sets are complete since data is accessed via SPARQL only. As long as endpoints do not support RDFS and OWL entailment, or as long as not all inferences are materialized, some classes may be missing.

Instead of inducing domain and range restrictions from RDF data, restrictions can also be induced from unstructured data. In [3], Cimiano et al. derive the classes of arguments of verbs from natural language text, which can be seen as a subtask in the context of ontology learning from text. Since in their work verbs are interpreted as binary relations, what they induce are domain and range restrictions. Given an existing taxonomy, for the domain and the range of a property the appropriate level in the taxonomy is selected considering the classes' conditional probabilities. Interestingly, the authors note that *"the domain and range of a relation can actually not be regarded as independent from each other,"* which is exactly what we do in this paper. However, due to a lack of training data, they refrain from regarding coupled domain/range restrictions.

In [8], Töpper et al. induce domain and range restrictions as well as class disjointness axioms from RDF data for the purpose of enabling to detect logical inconsistencies via reasoning. In their work, the domain (range) of a property is the class that most of the subjects (objects) in triples with this property belong to. This has the drawback that if for a property entities belonging to several diverse classes appear in subject (object) position, then the induced domain (range) restriction only regards the most specific superclass of these classes. While this is not wrong, for Natural Language Processing more fine-grained domain and range restrictions are interesting.

7 Conclusion

In this paper we presented the concept of coupled domain/range restrictions and presented an approach to apply Frequent Itemset Mining to induce independent domain and range restrictions as well as coupled domain/range restrictions from an RDF graph. Experiments carried out with the DBpedia dataset showed that high precision can be achieved. We believe that these restriction statements can be beneficially applied in Natural Language Processing scenarios such as Semantic Parsing, Question Answering, and Entity Classification. Therefore, all data obtained as well as our implementation is available to the community on a dedicated website.

Acknowledgements. This work was supported by the Cluster of Excellence Cognitive Interaction Technology 'CITEC' (EXC 277) at Bielefeld University, which is funded by the German Research Foundation (DFG).

References

1. Auer, S., Bizer, C., Kobilarov, G., Lehmann, J., Cyganiak, R., Ives, Z.: DBpedia: a nucleus for a web of open data. In: Aberer, K., et al. (eds.) ASWC/ISWC -2007. LNCS, vol. 4825, pp. 722–735. Springer, Heidelberg (2007). doi:10.1007/978-3-540-76298-0_52
2. Borgelt, C.: Frequent item set mining. Wiley Interdisc. Rev. Data Mining Knowl. Discov. **2**(6), 437–456 (2012)
3. Cimiano, P., Hartung, M., Ratsch, E.: Finding the appropriate generalization level for binary ontological relations extracted from the genia corpus. In: 5th International Conference on Language Resources and Evaluation (LREC 2006), pp. 161–169. Citeseer (2006)
4. Fleischhacker, D., Völker, J.: Inductive learning of disjointness axioms. In: Meersman, R., et al. (eds.) OTM 2011. LNCS, vol. 7045, pp. 680–697. Springer, Heidelberg (2011). doi:10.1007/978-3-642-25106-1_20
5. Fleischhacker, D., Völker, J., Stuckenschmidt, H.: Mining RDF data for property axioms. In: Meersman, R., et al. (eds.) OTM 2012. LNCS, vol. 7566, pp. 718–735. Springer, Heidelberg (2012). doi:10.1007/978-3-642-33615-7_18
6. Fournier-Viger, P., Gomariz, A., Gueniche, T., Soltani, A., Wu, C.-W., Tseng, V.S., et al.: SPMF: a Java open-source pattern mining library. J. Mach. Learn. Res. **15**(1), 3389–3393 (2014)
7. Leskovec, J., Rajaraman, A., Ullman, J.D.: Mining of Massive Datasets, 2nd edn. Cambridge University Press, New York (2014)
8. Töpper, G., Knuth, M., Sack, H.: DBpedia ontology enrichment for inconsistency detection. In: 8th International Conference on Semantic Systems 2012 (i-Semantics 2012), pp. 33–40. ACM (2012)
9. Völker, J., Niepert, M.: Statistical schema induction. In: Antoniou, G., Grobelnik, M., Simperl, E., Parsia, B., Plexousakis, D., Leenheer, P., Pan, J. (eds.) ESWC 2011. LNCS, vol. 6643, pp. 124–138. Springer, Heidelberg (2011). doi:10.1007/978-3-642-21034-1_9

Wikipedia and DBpedia for Media - Managing Audiovisual Resources in Their Semantic Context

Jean-Pierre Evain[1(✉)], Mike Matton[2], and Tormod Vaervagen[3]

[1] European Broadcasting Union (EBU), Grand-Saconnex, Switzerland
evain@ebu.ch
[2] VRT, Brussels, Belgium
mike.matton@vrt.be
[3] NRK, Oslo, Norway
Tormod.Varvagen@nrk.no

Abstract. The EBU, NRK and VRT are three European media companies. The EBU is the largest association of broadcasters. The NRK and VRT are the national public broadcasters in Norway and in Belgium (Flemish). The EBU, NRK and VRT are known in the media community for striving innovation. They have developed recognised expertise in engineering solutions and standards around the management of information for the audiovisual industry in a multi-lingual environment. They promote the use of semantic technologies for the production and distribution of content across a variety of media and platforms. In this context, Wikipedia, DBpedia, automatic metadata extraction and other tools are important information sources. This is not an academic paper but a report on the operational use of such information (access, usability, long term availability of information, editorial quality) and needs by the media industry. For broadcasters, minimizing cost and complexity is of the essence!

Keywords: Sport · News · Audiovisual · Semantic · Enrichment · DBpedia · Wikipedia · LOD

1 Introduction

As national public service broadcasters, NRK[1] and VRT[2] create and manage a vast amount of television and radio content distributed through linear or non-linear[3] channels to a variety of platforms and devices. The EBU[4] operates the Eurovision and Euroradio networks, which are important sources of entertainment, news and sport content for its 73 members over 56 countries. EBU also federates research among its members and all the knowledge and experience gathered in this paper is publicly shared within the EBU community.

---

[1] https://www.nrk.no/.
[2] http://www.vrt.be/.
[3] Providing access to content independently from a time schedule e.g. on demand.
[4] http://www.ebu.ch/home.

© Springer International Publishing AG 2017
M. van Erp et al. (Eds.): ISWC 2016 Workshops, LNCS 10579, pp. 41–56, 2017.
https://doi.org/10.1007/978-3-319-68723-0_4

Metadata has been neglected for a long time but it is now a first class citizen and an essential component of media digital production and distribution. Media companies multiply opportunities to gather and manage more information. Automatic metadata extraction from picture, sound and associated resources like subtitle provides a lot of additional data at an affordable cost (using NLP for entity recognition, primarily persons, organizations, locations, events). In a semantic framework, the collection of Linked Open Data from Wikipedia and DBpedia[5], among others, is another complementary source of data, in particular when developing knowledge bases on the entities mentioned above. This is already largely exploited but more could be done for more specific data like sport results, exploiting more than e.g. the Wikipedia infobox.

The present paper reports on developments and operational usage of LOD in the media industry from sources like Wikipedia, DBpedia and other sources. It also highlights new expectations and proposes changes or new approaches to improve access and usability of this information in a multi-lingual environment. Even if the perception of "state-of-the-art" is understandably different in the academic community, We can say the following is "state-of-the-art" for operational services. Sharing our experience is about filling the gap between system integration and academic innovation.

Beyond the technology itself, it is also expected that this will facilitate crowdsourcing of important information of high editorial quality made publicly available.

2   Using Semantic Data at the EBU

2.1   Engaged in Semantic Web and Linked Open Data

The EBU has been promoting the use of semantic technologies in broadcasting production and distribution for around 10 years. Recently, the EBU has developed its collaboration with organizations like the International Olympic Committee, the Movie Labs (the joint research laboratories of the movie studios in North-America), and several broadcasters. The EBU is also chairing an activity around semantic modelling of metadata for service oriented production architectures[6] in the international FIMS project[7]. This shows a real trend in the adoption of semantic technologies in the media industry.

In this context, some of the primary benefits of semantic modelling perceived by broadcasters are flexibility and scalability. The "triple" is seen as a new ultimate common metadata format in which can be converted information from production and distribution data silos. But facilitating the aggregation of additional richer Linked Open Data is also well perceived as another significant advantage.

The focus of the EBU semantic developments and specifications is around audiovisual content archives as well as news and sport live exchanges. The most common set of entities is persons, organizations, locations and events. The model also supports publication events and services like in schema.org (supported by Google, Bing, Yahoo), which was extended in a joint contribution from EBU and BBC. Wikipedia, DBpedia

---

[5] http://wiki.DBpedia.org/.
[6] SOAP and REST.
[7] Framework for Interoperable Media Services, http://www.fims.tv.

and other repositories are a mine of information to enrich a knowledge base suitable to media needs. DBpedia Such knowledge bases can also be used to illustrate common concepts like news topics, etc.

What could be done to get even more?

2.2 EBU's Usages for Wikipedia/DBpedia

The EBU has developed the EBUCore,[8,9] EBUSport and Conceptual Class Data Model[10,11] ontologies.

EBUCore is publicly accessible and published as a Linked Open Vocabulary (LOV[12]). EBUCore provides a semantic framework for managing audio-visual resources and business applications such as news.

The EBUSport ontology (Fig. 1) builds upon EBUCore to manage and link all sport-related data (e.g. competition events, start lists, results) to footage. EBUSport is derived from the Olympic Data Feed XML schemas provided by the International Olympic Committee (IOC)[13]. The EBU has been contracted by the IOC to work on the mapping of ODF sport results into RDF.

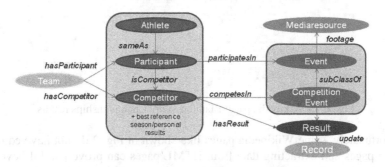

Fig. 1. EBUSport model in a nutshell

Data is aggregated from Wikipedia, DBpedia, sport federation web services or sport event data streams and integrated in the dataset using EBUCore and EBUSport properties.

2.3 DBpedia Data Resources in EBU's Sport Applications

EBU acquires sports rights and covers majors sport events (UEFA, FIFA, Biathlon International Federation, IAAF for athletics, Cycling with "Tour de France", tennis, and

---

[8] http://www.ebu.ch/metadata/ontologies/ebucore/ (html documentation).

[9] http://www.ebu.ch/metadata/ontologies/ebucore/ebucore.rdf.

[10] The CCDM is used to model content production, management and distribution workflows.

[11] https://tech.ebu.ch/docs/tech/tech3351.pdf.

[12] http://lov.okfn.org/dataset/lov/.

[13] http://odf.olympictech.org/ (International Olympic Committee – Olympic Data Feed).

many others) to provide content to its members via the Eurovision[14] network. It also provides services like the Worldwide distribution of the Olympic Broadcast Service's Multi-feed Distribution Signal, the ONC (Olympic news channel) and ad-hoc unilaterals like in Rio 2016.

A lot of metadata is being exchanged in preparation or in the course of a sport event. Information is shared via sport federations, athletes websites, national delegations, event hosts, companies measuring performances and relaying results. It is all being used to enrich our knowledge base transforming e.g. CSV, pdf or XML data into RDF[15].

We also fetch additional data from Linked Open Data sources like Wikipedia and DBpedia or sport federation web services, typically about the entities mentioned above but also information about past events like results shown in Fig. 2.

Fig. 2. A Wikipedia page: Biathlon WorldChampionships results

Unfortunately, some Wikipedia pages like shown in Fig. 2 do not have equivalent DBpedia pages and extracting data from HTML pages can prove painful beyond the infobox. Additionally, the structure of these pages may vary significantly prohibiting efficient automatic parsing and data extraction.

Natural Language Processing (NLP) should help in the translation of more legacy Wikipedia information into DBpedia. Such techniques are used in the media industry e.g. using entity recognition or other knowledge extraction after speech to text conversion. But as far as Wikipedia is concerned, the media industry would rather directly use DBpedia resources and the better if initially further enriched from Wikipedia using NLP. Broadcasters would not use NLP to extract data from Wikipedia because of the inherent cost and complexity of integration in their data workflows.

Schema.org is not an option as it doesn't represent an alternative to Wikipedia or DBpedia. The query domain would be too wide. Schema.org has been designed for data to be harvested and indexed by search engines, which is beyond the operational capabilities of broadcasters.

---

[14] http://www.eurovision.net/.

[15] http://www.w3.org/TR/2014/NOTE-rdf11-primer-20140624/ (RDF 1.1 Primer).

Parsing DBpedia's RDF files is mostly what we are doing today to enrich e.g. athlete profiles as shown on Figs. 3 and 4. We start from a list of athletes, generate the URLs, access and parse the RDF files with XSLT (rather than querying the SPARQL endpoint), extracting and mapping the data from a pre-selected set of properties, automatically. Going over 2,500 athletes takes less than 10 min. The results are reasonably accurate and reliable (less than 5% of non-athlete profiles have been inadvertently ingested). This is in fact another simple and effective way of using NLP in the sense of using natural language as available.

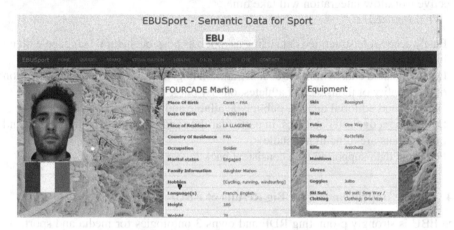

Fig. 3. Upper part of an athlete page in the EBUSport application

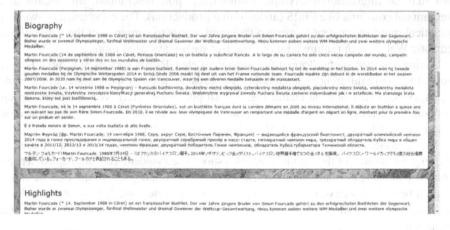

Fig. 4. Lower part of an athlete page in the EBUSport application

For that reason, Wikidata is not seen as a mature solution (yet). Abstracting all resources and properties behind keys requires additional human lookup, hence operational complexity and cost.

There is unfortunately no resource for sport like IMDb[16] around movie and TV celebrities, which is anyway not an open resource either.

Other databases exist addressing a variety of concepts like geonames[17] for "countries" with associated geolocation. However, finding a comprehensive resource, preferably publicly available free of charge, remains a challenge. When available, parsing websites, web services and various data formats is time and resource consuming.

Of course, broadcasters use automatic extraction tools such as speech to text associated to NLP techniques for entity recognition. But here too, cost is of the essence and effective workflow integration will take time.

What we need is:

– Easily accessible structured data using as much as possible natural language with no intermediary look-up step,
– Disambiguated data to limit the cost of curation (e.g. avoiding the human verification of the identity of thousands of athletes, one by one),
– Data of good editorial quality, publicly available and preferably free of charge,
– Data with recognised datatypes, in particular date and time (not recognised date and time formats are difficult to handle),
– Persistent data supported by a consistent backup and update strategy.

2.4   Simple Improvements for Big Results at EBU

The EBU is strongly promoting RDF and owns 3 ontologies for media and sport. As already mentioned, the EBU is sharing expertise with its members and other organizations like the International Olympic Committee, sport federations, MovieLabs and international consortia like FIMS (audio-visual production). The experience shows that a practical proof of concept is worth a thousand words to convince of the value of the Semantic Web, which require a lot of data. But we always need more…

Having a unified syntax/format for person/athlete URLs is very useful. But it is well known that many different characters or structures are used to spell names in different languages (special language characters, apostrophe, dash, etc.). We recommend that Wikipedia and DBpedia propose best practices for character/structure mappings, or explain better what tools are available to implement these mappings.

As mentioned above, disambiguation is vital. Data curation is costly and manually verifying the identity of thousands of athletes should be avoided. The language of a page is one key filtering attribute but more can be done on homonyms. In the case of sport, it would be more effective to use two parameters like "role" and "sport" (e.g. "athlete" and "biathlon" instead of "biathlete" which combines the two notions of athlete and discipline/sport).

Well known and defined datatypes (in particular for date and time) should be used to facilitate further conversion in different development environments like e.g. Json and Javascript.

---

[16] http://www.imdb.com/.
[17] http://www.geonames.org.

It is well understood that DBpedia has to deal with a lot of Wikipedia legacy. It is also clear that it is easier for Wikipedia authors to generate HTML pages rather than populate a DBpedia list of properties. However, we believe the future should give the priority to DBpedia as the reference, which data would be used to generate Wikipedia pages. DBpedia HTML templates could be solution for gathering data to be published as today keeping the RDF files as a key export format.

For Wikipedia, the examples of athlete pages show that there is a lack of structure beyond the infobox. The page body often contains "subjective"[18] content that even NLP would not be able to effectively sort out. We would recommend providing more guidance on themes to be addressed and a structure to host them.

Similarly, templates should be provided for Wikipedia tables. Without it, tables are difficult to process and require advanced programming (using e.g. JSoup). The EBU would propose properties for sport data tables in order to facilitate the publication of such pages in DBpedia (which is not the case today).

The availability of RDF files is certainly a "must". More would be welcome! The EBU would be happy to share part of its ontologies to provide a set of additional properties for audio-visual content (including links to YouTube) and sport data.

3 Using Semantic Data at the NRK

3.1 Changing the Role of Metadata

In the last two years, the role of metadata has changed a great deal in NRK. Previously, metadata was mainly used for retrieval of archive material, and the metadata was added at the end of the material life cycle for this purpose. The architecture and file-based workflows did not dramatically change from the tape-based infrastructure it was designed to supersede. After ten years of file-based production, the old Media Asset Management (MAM) system had to be replaced. This time we gave us the opportunity and time to rewrite the whole architecture from scratch.

In the former infrastructure metadata was stored in different domain-based relational databases. The data model was vendor-driven. Each system we bought had a different data model. Metadata was very fragmented and a huge part of it got lost during the production process.

The new architecture shown in Fig. 5 is based on one common RDF based metadata layer, based on EBU CCDM[19] and EBUCore, connecting all the systems as well as storing all the different states and versions of metadata describing programmes and parts of programmes, produced during the entire life cycle of content production.

Radio, television and the web are now using the same content base. Data fragmentation between systems has been solved thanks to the use of semantic technology and new services supporting new ways of storytelling.

---

[18] Wikipedia authors can express non-factual views, which may potentially affect the editorial quality of an article or the quality of the data being extracted.

[19] https://tech.ebu.ch/docs/tech/tech3351.pdf (EBU Class Conceptual Data Model).

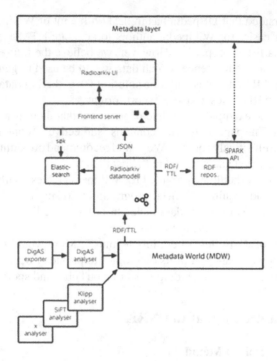

Fig. 5. New NRK metadata architecture

3.2   Web First

One year ago the CIO of NRK decided that web-based content distribution should be given higher priority. Looking at consumption statistics, online audio/video publishing went from being a third priority service, mainly a catch-up service for the linear channels and short clips in web articles, to a first priority core business. At this point some of the programmes already had more viewers online than the traditional linear broadcast distribution. The trend started with children's programmes. Children adopt new services naturally, and for them a television set is older technology than an iPad. The ongoing trend is spreading to more genres of programmes and to a variety of age groups.

The web is always ready for new content; you never have to wait for airtime nor break into another show. That makes it the fastest media for breaking news and other events like sports. We still use the linear channels, but web-based content is the most important and getting more popular.

Online publishing gives us more possibilities for additional content, facts and background information, and the audience expect and demands such supplementary content changing what used to be the traditional story told. Meanwhile, the number of journalists is not increasing. Support staff like researchers and librarians is decreasing and kept to a minimum.

In this context, the extraction of background material and facts is vital, both as part of the research, and for automatic generation of suitable background facts published as a part of the NRK branded content.

Semantic technologies help tremendously in providing good quality content and metadata, covering sport, news and current affairs. It is imperative that same high level of precision and editorial quality is kept for supplementary content as for the content produced by the journalists. Any mistakes here will not be tolerated by the audience.

3.3   Authority Registers

One of the first tasks we started once the new data model was in place was to establish multiple authority registers, also referred to as the knowledge base. It was found that we had to have better control of places, persons, events used in the day to day production. First task was to make a register of place names, everything from countries and towns, to recording places and concert venues. We took the basic list of place names from our online weather service (http://yr.no) and added new places and additional language variations from multiple sources. One of the challenges is that placenames have different spelling in different languages, and sometimes it can be confusing which place the news story are referring. Here we got good help from sources like geonames.org and DBpedia.org.

Another challenge is the multiple occurrence of the same place name for up to 20 different locations in the country, the map coordinates are of course not the same, but it makes it hard to use interfaces like text based auto complete.

The weather service was already multilingual, and we enhanced the data with all the language variations we could find and the data we linked was mostly of good quality. It is rather simple data models we have here, but still some manual labour was needed in the cleaning and disambiguation processes.

The next thing we gathered was a list of persons related to the content, as part of the metadata driven production infrastructure project. The list contains composers, artists, talents, politicians, technical staff, and even people mentioned in a programme. The identity of contributors is often used for linking programmes together. Another usage is reporting of use of intellectual rights, where unambiguously identifying right holders is of paramount importance. It was a much more complex task working on persons than working on locations. Data from existing archives had to be merged, and enhanced from external sources. We had again problems with duplicates (several persons sharing the same name). Like for the EBUSport example above disambiguation was key but we very often ended up with manual tasks to connect to the right person. Furthermore, our register has a broader scope than sport, making the problem bigger.

Example: handling music publishing rights requires documenting the death date for each composer, because intellectual works that are free to use 70 years after the creator has died. This information can be found on Wikipedia or DBpedia when the mapping was possible. In most cases, this information will be accessible from Wikipedia, but on certain occasions, even if present, the information is buried in the text and it is not possible to extract this information automatically (incl. for mapping to DBpedia). It then becomes an expensive and time consuming task to curate data and determine where the information is missing or incorrect in the data set. Of course in many cases changing the language will help, but for lesser known persons, this is not an option.

Right now the collected data (places, persons, etc.) is a good tool for research and provides a solid basis for daily journalistic work. But it is still not good enough for automatic publishing of supplemental material incl. data journalism. We still need the editorial approval from the journalist before publishing, which means that the journalist must have full knowledge about the topic. This is sadly not always the case. If we can't trust the source 100%, if we can't control the quality, then we can't publish the content at all.

We would also benefit from a more systematic access to a common core set of mandatory properties either in the infobox or in more structured body text. NRK can suggest properties that would relevant to media such as systematic geo-location of places or key dates.

3.4 Pros and Cons of Using Wikipedia and DBpedia

Wikipedia is a wonderful collection of articles in multiple languages through which everyone can easily share knowledge. The author can start from a blank sheet and the system even proposes some formatting tools. Easy editing was the key to success and probably the only way to get it up and running at the start. It has grown now to a humongous number of articles, but sadly not sufficiently structured for today's needs, in other words "automatic metadata extraction by machines". DBpedia is a partial answer but a significant amount of information is missing or many Wikipedia pages simply do not have their DBpedia equivalent.

Another problem is the discrepancy of information across national version of Wikipedia. Let's take the page for the "Goldberg Variations" as an example. The page in English is richly documented and has even links to score and recordings. But the name of the parts of the work are in the headings above each chapter where in the German version uses a proper table, but have less content in the sound and score departments. In comparison, the Norwegian page has hardly any useful information beyond what users would already know. Fortunately, for other topics, the Norwegian DBpedia can be an excellent source of information.

3.5 Simple Improvements for Big Results at NRK

More rules and instructions indicating how data should be presented in tables or structured Wikipedia pages (e.g. more specifically marked-up HTML) would help re-using published information. Data and facts need structure plus good metadata for being useful and connectable. NLP techniques can help, but facts in the main text are best read by people, not machines. Data extraction using NLP may generate "noise", burying good data even on DBpedia pages. It is however understood that this is a possible temporary solution for legacy content.

Wikipedia and DBpedia are well installed. Now might be right time to give the priority to DBpedia as the source of information to populate structured Wikipedia page, Wikipedia being just the "renderer" for human readers. This of course means new

DBpedia templates and other editor friendly tools as well DBpedia to Wikipedia publication gadgets.

Such a change in direction like this will result in higher data quality and will make Wikipedia and DBpedia a set of reliable information sources for broadcasters in the future. This would also help broadcasters share more information in return.

4 Using Semantic Data at the VRT

4.1 The Scope Shift in Descriptive Metadata

Within a broadcast organization such as VRT, metadata is of crucial importance to drive business processes. First of all, correct technical metadata about audiovisual content creates interoperability between different production systems, and most system integrations focus a lot on this technical metadata interoperability. Descriptive metadata is often overlooked in system integration.

Broadcasting and media production are currently undergoing transformations, and this descriptive metadata, i.e. metadata that describes what the audiovisual content is about, is becoming ever more important. It describes what is appearing in the video, or what can be heard on the audio. Traditionally, descriptive metadata has always been seen as the domain of the archives. Content was archived together with a description in order to more easily retrieve it in a later stage. This was mainly due to the linear production and distribution process of TV content, in which there was not really a need for descriptive metadata during the production process itself. The people that were producing it knew well enough.

Things have changed and the rise of non-linear content through all kinds of distribution platforms has led to an explosion of content that needs to be published. E.g. news stories need not only be distributed on linear TV, but also need to be available in an itemized way through various on-line platforms and through social media, on different devices: TV, laptop, smartphone, tablet, etc. Furthermore, as the amount of content explodes, consumers more and more expect to be served in a personalized way in order to find the content of interest to the user. Providing each of these platforms with the right kind of content in a manual way is plainly impossible as resources in media organizations are limited. Lastly, also the amount of audiovisual content is ever increasing, with users becoming more and more interactive and generating a lot of user generated content.

Therefore, there is a constant search for methods to automate the processes of delivering the right kind of content through the right kind of platform to the right target audience. Structured and affordable descriptive (and technical) metadata is of crucial importance to make this possible, and Semantic Web technologies enable natural ways to represent and exchange knowledge about (audiovisual) assets.

4.2   Linking Descriptive Metadata to the Semantic Web

VRT has been exploring linked open data for several years, with several R&D projects starting in 2010. The main motivations for considering the use of LOD are:

More structured information. As more and more business processes in media companies need to be automated, we have a need for structured information, and this includes structured descriptive metadata. Semantic Web languages create a natural way for storing such structured information.

Enrichment. Another reason for starting to work with Semantic Web is the potential to link your own knowledge, with world knowledge through connecting the concepts in your own ontology with LOD sources.

Increased interoperability. Furthermore, as information in LOD is always structured around triples (as opposed to relational databases which have highly irregular structures), it is easier to interconnect systems and to make data interoperable.

Multilingual linking by nature. Finally, media companies deal a lot with content in different languages. Many LOD sources have a natural way to link concepts together independent of the language. Examples are DBpedia, BabelNet[20] and Wikidata[21].

4.3   Case Study 1: Semantic Search Engine (MediaLoep)

The first explorations of VRT with Semantic Web technology to support content retrieval happened several years ago in the form of a search engine empowered with Semantic Web technology. The full MediaLoep system contains the following components:

1. A triple store (OpenLink Virtuoso)
2. Enrichment services connected to Linked Open Data sources
3. A search engine (to allow more efficient searching)
4. A semantic-powered user interface

An ontology has been created on top of a simple media production data model, consisting of an Editorial object with descriptive information, and a MediaObject to which time-based annotation can be added (e.g. subtitles). This simple model is shown in Fig. 6. This simple model was further extended to support different content types (broadcast news, fiction, etc.). A full description of the model can be found in other publications for a more detailed description [1, 2].

Additionally, all descriptive metadata available in the different production systems at VRT was combined and attached to this data model for each content item. Eventually, our ontology allowed us to make all data and annotations available in RDF in the triple store.

---

[20] http://babelnet.org/.
[21] https://www.Wikidata.org/wiki/Wikidata:Main_Page.

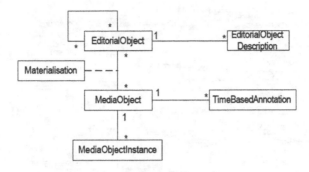

Fig. 6. The basic VRT data model close to EBUCore

One particular information source is a thesaurus with keywords. Each content item in the archive is tagged with some relevant keywords for this item. Within the MediaLoep ontology, this thesaurus has been implemented using SKOS[22]. An excerpt from the full ontology is shown in Fig. 7. This thesaurus, containing place names, person names, and other keywords such as events, topics etc., is an ideal candidate to link to other linked open data sources. The main sources with which the thesaurus terms have been linked are DBpedia and GeoNames. Linking to only those 2 data sources already allowed for plenty of useful features to augment the semantic-enabled search engine MediaLoep with:

- Automatic mapping of content items to locations on a map
- Map-based querying of the content archive
- Automatically adding profile pictures to people
- Allowing semantic-enriched queries (e.g. I want to retrieve all content items with president-predecessors of Barack Obama).
- Automatically include synonyms of search terms in the query
- Etc.

Explaining the whole system would justify a full publication in itself us too far for this paper, but we propose a link to a youtube video for a quick overview of the features proposed in MediaLoep[23].

---

[22] http://www.w3.org/TR/skos-primer/.
[23] https://www.youtube.com/watch?v=CIaLnLKuXu8.

Fig. 7. An eagle-eye view of the MediaLoep ontology

4.4 Simple Improvements for Big Results at VRT

In order to be even more useful for media companies to work with LOD, the following additional functionalities, or structured data, would be very useful.

Placing semantic concepts in a time context. Many semantic concepts have relations that potentially vary over time. This is for instance true for persons, which have different roles through their life. E.g., depending on the time context, the concept "President of the United States" is referring to a different person. Within the current structure of LOD (and DBpedia in particular), it is quite hard to retrieve the correct "version" of the semantic concept given the time context. Another example is sports teams such as soccer teams, where athletes join and leave over time. If we have media content about a particular soccer team, and we know the recording time of the media content, it is now not trivial to enrich the media content with the list of soccer players that were active at that time. We suggest studying possible ways to improve this situation, by also structuring the time concept.

Furthermore, versioning and provenance information about semantic concepts and information is of importance, but as research on these topics is ongoing, it is only a matter of time before useful solutions arise.

Dumps or online "live" data. Another critical point for operational use of LOD is the long-term availability of the data source. In order to keep operational support under control, we have been copying the required knowledge from linked open data sources into our own ontology. The disadvantage of this way of working is that we miss live updates into our knowledge repository, such as new places, new persons, new relations of existing persons etc. On the other hand, our company misses the required resources to persistently monitor updates.

Also, entire data sources might change. An example of this is the recent move from freebase, which will be shutting down, to Wikidata. If services in the media company heavily rely on freebase, it will take quite some effort to migrate the existing systems to make use of Wikidata instead. For those reasons, copying the required information will probably remain popular. However, it would be really useful if there are ways to automatically receive triggers about "updates" to the semantic concepts you have in use, which would allow to more easily correct errors or integrate new knowledge into your knowledge repository. Even if recent advances on this topic are improving the situation, updating knowledge in live production systems is a risky business if these production systems rely on it, and fully automated testing before committing updates is not (yet) in place. This last point however is only a matter of resources to tackle it.

5 Conclusions

This paper shows the important role that Linked Open Data resources like Wikipedia, DBpedia, or other data sources plays for the media companies. The EBU, NRK and VRT have presented their development an operational work, which is representative of what is currently being made in the broadcasting industry, even if significantly and naturally behind academic research.

In order to close the gap between research and operation, the requirements and needs expressed in this paper propose a number of reasonable changes, which, will significantly facilitate adoption of Wikipedia, DBpedia and other linked data services by the media industry if they are put in practice.

As far as NLP is concerned, this is well understood the media industry already makes an intensive use of such techniques after automatic metadata extraction like Named Entity Recognition, speech to text or text to speech. However, we see NLP as a possible way to extract knowledge from existing unstructured pages, but broadcasters will not use it directly (too expansive and expensive). The results will be used by broadcasters if brought back to DBpedia or Wikidata.

We would also suggest that DBpedia or Wikidata takes a different role and becomes the reference in the form of structured data, being used for subsequent publication by Wikipedia or for other applications. EBU, NRK and VRT are looking forward to share more common classes and properties to help publishing more structured data as e.g. RDF in DBpedia.

Of course, there are advanced state-of-the-art NLP techniques. But the question is "why do we seem to be unaware of these tools or ignore them?". We actually know they exist but they remain hard to integrate in operational workflows. We do support research but our everyday need is primarily around affordable and implementable solutions. Cost and simplicity is of the essence!

Acknowledgements. The authors would like to thank EBU, NRK and VRT for authorizing them to write this paper and report on recent Research & Development as well as operational business-related activities.

The authors also wish to thank the Wikipedia and DBpedia communities, and supporting teams, for the tremendous work done.

References

1. Debevere, P., Van Deursen, D., Van Rijsselbergen, D., Mannens, E., Matton, M., De Sutter, R., Van de Walle, R.: Enabling semantic search in a news production environment. In: Proceedings of the 5th International Conference on Semantic and Digital Media Technologies, Saarbrucken, Germany, December 2010
2. Debevere, P., Van Deursen, D., Mannens, E., Van de Walle, R., Braeckman, K., DeSutter, R.: Linking thesauri to the linked open data cloud for improved media retrieval. In: Proceedings of the 12th International Workshop on Image Analysis for Multimedia Interactive Services, Delft, The Netherlands, April 2011

Identifying Global Representative Classes of DBpedia Ontology Through Multilingual Analysis: A Rank Aggregation Approach

Eun-kyung Kim[✉] and Key-Sun Choi

School of Computing, Korea Advanced Institute of Science
and Technology (KAIST), Daejeon, Republic of Korea
{kekeeo,kschoi}@world.kaist.ac.kr

Abstract. Identifying the global representative parts from the multilingual pivotal ontology is important for integrating local language resources into Linked Data. We present a novel method of identifying global representative classes of DBpedia ontology based on the collective popularity, calculated by the aggregation of ranking orders from Wikipedia's local language editions. We publish the contents of this paper on http://semanticweb.kaist.ac.kr/home/index.php/DBBO.

Keywords: DBpedia · Multilingual · Ontology · Rank aggregation

1 Introduction

The diversity and amount of data on the Web are both continuously growing, and there has been a paradigm shift leading from the publishing of isolated data to the publishing of interlinked data through a variety of knowledge sources such as Linked Open Data (LOD) [1]. DBpedia dataset [2] currently plays a central role in the LOD cloud, which has been populated using a large amount of collaboratively edited material (i.e., Wikipedia) as a knowledge source. Because of the ever-growing size and enormous scope of Wikipedia's coverage, the DBpedia dataset has been increasingly applied to a wide range of web applications.

The DBpedia dataset contains a community-curated cross-domain ontology to homogenize the description of information in the knowledge base (KB), which is one of the largest multilingual ontologies developed to date. Version 2014 of this ontology covers 685 classes in total, which form a subsumption hierarchy, and includes 2,795 different properties. This ontology has become a *de facto* reference vocabulary; however, this is limited as a multilingual pivot. Although a large number of instances among different languages are connected to the owl:sameAs[1] link, matching the class level is rare. The rdfs:label properties use language tagging to enhance multilingualism as follows.

---

[1] https://www.w3.org/TR/owl-ref/#sameAs-def.

© Springer International Publishing AG 2017
M. van Erp et al. (Eds.): ISWC 2016 Workshops, LNCS 10579, pp. 57–65, 2017.
https://doi.org/10.1007/978-3-319-68723-0_5

```
<owl:Class rdf:about="http://dbpedia.org/ontology/Actor">
    <rdfs:label xml:lang="en">actor</rdfs:label>
    <rdfs:label xml:lang="fr">acteur</rdfs:label>
    <rdfs:label xml:lang="ja">俳優</rdfs:label>
    <rdfs:label xml:lang="ko">영화인</rdfs:label>
    ...
```

This shows that the class "Actor" has several cross-lingual corresponding terms such as " 영화인 " in Korean and "Acteur" in French. Figure 1 shows the statistics of class numbers with `rdfs:label` properties. The number of labeled classes for different languages varies significantly, and there is obviously an absence of cross-lingual labeling for some editions such as Chinese. The DBpedia ontology (DBO) is continuously evolving due to its collaborative (wiki) paradigm and ongoing internationalization [3,4]. However, it suffers from a scarcity of multi-lingual labels, due to its derivation that is based on the popular infoboxes in English. This leads to a limitation of other languages' ability to adapt the DBO to local language knowledge resources and makes it difficult to homogenize as a conceptual extension. Thus, identifying the global representative parts of the DBO is important for expanding multilingual ontologized space in LOD.

Figure 2 gives an overview of our motivation. Generally, the terminological components (henceforth referred to as the TBox) of an existent ontology can be translated and tailored to fit the understanding of other languages to expand multilingual coverage and thus increase knowledge access across languages with existing ontologies [5]. Therefore, a multilingual pivotal ontology must accurately

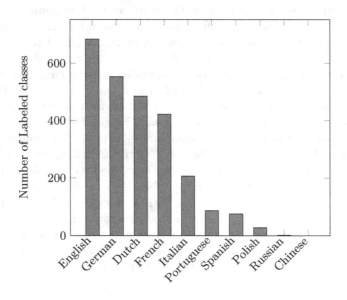

Fig. 1. Statistics for language-labeled classes of DBpedia ontology among 10 major languages

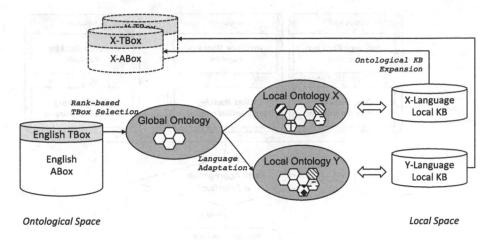

Fig. 2. Multilingual ontologized space expansion

represent the global common concept structure, yet remain reusable in different languages so that connections can be made between local language knowledge resources and ontological KBs when entering an LOD.

We aimed to identify globally representative DBO classes for different language editions in this work, based on the combination of several ranking results that analyze the knowledge graph to measure the popularity of instances from multiple perspectives. Then, a consensus global ranking could be produced via rank aggregation; finally, we constructed a representative subset of DBO that could capture universally popular information that would be useful for improving the multilingual reuse of the ontology itself and would more easily and rapidly expand the ontological domain of the local language knowledge sources. We evaluated our approach by comparing its coverage with respect to the losses caused by the selection process, which had almost the same coverage with no appreciable loss of efficiency for larger sizes when the data were adapted to multilingual purposes.

2 Rank Aggregation–Based Class Selection

When determining globally representative classes of DBO, we believe that the main challenge lies in the ranking model. Figure 3 shows an overview of the proposed approach that is mainly structured as two phases, as in the following subsections.

2.1 Language-Specific Popularity Analysis

We create a ranking model of classes to ascertain their degrees of significance in the ontology by analyzing each language dataset individually. We first computed the ranking order for its instances and then combined these to determine the

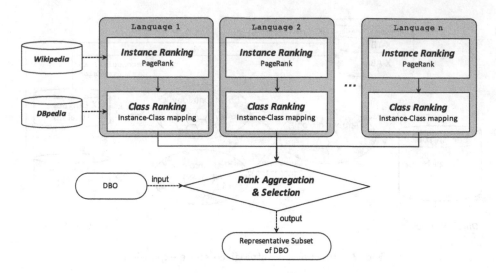

Fig. 3. Overview of the proposed framework

rank of a class. The rank of a class for each specific language is obtained by the PageRank [6,7] values of its instances. We constructed a graph of instances from Wikipedia consisting of the links between articles to calculate the ranks of the instances. Each article corresponds to a node of the graph, and links between articles correspond to the edges of the graph.

Then, we calculated the rank of a class by mapping information between the instances in Wikipedia and the classes in DBpedia. We used "type" information from DBpedia to map instances to classes; for example, the instance "Barack Obama" is described and classified in three types as "OfficeHolder," "Person," and "Agent." Our class-level ranking model characterizes the following two features of a class to determine its rank:

1. A class is more popular if it is ranked higher based on the *average* of its instances' rank scores.
2. A class is more popular if it is widely *populated* in DBpedia ABox (i.e., the assertional component).

We used an aggregate function ($average * counting$) to compute the language specific class-level rank $CR^l(\text{C})$ of a class C as:

$$CR^l(\text{C}) = \frac{1}{n} \sum_{i=1}^{n} PR(i) * \sqrt{\|\text{C}\|}, \tag{1}$$

where n is the number of instances of C, $PR(i)$ is the PageRank score of instance i, and $\|\text{C}\|$ indicates the unique number of populated instances of C in DBpedia ABox for a language l. In the results, $CR^l(\text{C})$ represents the popularity of class C in the language l edition.

Table 1. Top 10 classes in different languages ranked by proposed approach; the distinct classes in each language are marked in bold type

English	French	Portuguese	Polish
Country	Country	Country	Country
Continent	**State**	Place	Place
Place	Continent	PopulatedPlace	PopulatedPlace
PopulatedPlace	**Department**	Agent	City
Agent	PopulatedPlace	Person	Agent
Organisation	Place	Organisation	Settlement
Person	Agent	Settlement	Person
Settlement	Person	City	Organisation
City	Settlement	**Artist**	**Region**
AmericanFootballTeam	**Territory**	**Work**	**AdministrativeRegion**

2.2 Language-Unified Popularity Analysis

The individual ranking orders from Sect. 2.1 are aggregated to produce a "globally popular" order of classes that would reflect their order of importance as judged by the collective evidence of all language editions. Table 1 depicts the top 10 independently ranked classes in four different language editions (the four sample languages in Fig. 1). This means that different language editions of DBpedia may have different perspectives on the information that they contain. We produced the consensus rank for each language-specific ranking order using the existing score-based rank aggregation method (i.e., the Borda count method [8]). The Borda count is one of the most well-known and intuitive rank aggregation schemes in which each element for each ranking order is given a score depending on its rank, and these weights are then summed across all such ranking orders.

Each language-specific ranking is associated with a finite set of m classes $C = \{C_1, ..., C_m\}$, each of which is given a score depending on its place in the individual ranking order, the Borda scores are summed for all such individual scores to compute their total score. More formally, each class C_i has a different ranked position x, which is based on the class ranking function $CR^l(C_i)$. We then define $\tau^{l_j}(C_i) = x$ ($1 \leq x \leq m, 1 \leq j \leq n$) such that the jth language edition ranks the class C_i at the xth position. Each class C_i has a Borda-based global ranking $CR^g(C_i)$, defined as:

$$CR^g(C_i) = \sum_{j=1}^{n}(m - \tau^{l_j}(C_i)). \qquad (2)$$

TBox Selection: After computing each class's global rank, the classes with the higher order global ranking scores are selected as the representative $\mathbb{R}$ with a certain size ρ by the following to define the classes and properties that should be included in:

Definition 1. *The set of classes $C(\mathbb{R})$ contains a class* C *iff:*

- C *is a class and* $CR^g(\text{C}) \geq CR^g(\text{C}_\rho)$ *or*
- C *is a class and there is a class* D $\in C(\mathbb{R})$ *such that* D $\sqsubseteq$ C

Definition 2. *The set of properties $P(\mathbb{R})$ contains a property* p *iff* p *is a property belonging to a class* C $\in C(\mathbb{R})$.

3   Experimental Analysis

We measured the coverage of the representative subset; good representatives are expected to capture most of the information in the initial dataset without much loss. We compared the performance of our algorithm with two others: Monolingual-Rank (Mono) and Random-Selection (Rand). Mono is an approach that uses only English to calculate the rank computation. Rand represents the average performance of 10 runs by randomly selecting a subset of the dataset as a representative.

We used the DBpedia Mapping-based Dataset (2014) in our evaluation, which is a set of assertion triples that contain very specific information about the entities that can be used to query Wikipedia. Every instance in those triples is classified by the classes of DBO, and all properties are defined in the ontology. A sample RDF statement (in triple form: $< s, p, o >$) of this dataset that pertains to "Barack Obama" is as follows.

```
PREFIX dbo: http://dbpedia.org/ontology/ PREFIX dbr:
http://dbpedia.org/resource/

<dbr:Barack_Obama, dbo:birthPlace, dbr:Hawaii>
```

This shows that the resource "Barack Obama" is the subject of other statements and presents a triple describing "Barack Obama"s birthplace as Hawaii.

We vary the number of representative subsets from 1 to 562 (the number of all classes involved in the rank; nearly 18% of the DBO's classes are never used in any languages) and compare the coverage achieved by the three methods listed in Table 2. It is clear from the results that the extracted globally popular classes have helped realize higher coverage for many languages.

3.1   Evaluations

We used gold standards[2] that were derived by assuming possible characteristics from both the number of the existing `rdfs:label` and the persistence of

---

[2] Evaluation data for this work is available for download at http://semanticweb.kaist. ac.kr/home/index.php/DBBO.

Table 2. Coverage of representative set for ten languages defined in Fig. 1. Percentages of triples are defined by classes in Ours, Mono, and Rand. |ℝ| represents the size of the selected classes

|ℝ|	Ours	Mono	Rand
1	27.85%	0.85%	0.03%
5	67.54%	39.69%	0.03%
10	80.90%	67.54%	0.03%
20	84.54%	83.99%	2.26%
50	91.39%	84.60%	6.57%
100	93.05%	91.61%	12.96%
200	94.27%	92.31%	27.51%
300	94.74%	94.56%	42.35%
400	96.86%	94.74%	57.98%
500	97.13%	-	72.30%
562	97.18%	-	83.02%

the classes. Then, we extracted a subset of DBO that could be adapted as the groundwork for automatic DBpedia mapping among languages through experimental evaluation.

For the first evaluation, we assumed that the classes that already have cross-lingual labels in many languages are important, because the labeling effort indicates their potential reuse in other languages. We create a gold standard based on this assumption by calculating "the number of existing `rdfs:labels` (Gold-standard 1)" for each class of DBO and divided the classes into two sets, positive and negative, in accordance with this assumption. (1) Positive set: Classes containing five (the mean value of DBO) or more labels. (2) Negative set: Classes containing fewer than five labels.

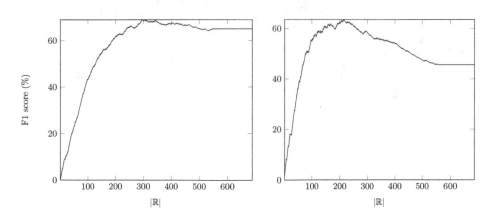

Fig. 4. F1 scores of Gold-standard 1 (the left side) and Gold-standard 2 (the right side) with respect to |ℝ|

For the other evaluation, we assumed that the classes that are preserved across the two versions (the first and latest DBOs) are more important than the newly added classes in the 2014 version called "persistence (Gold-standard 2)." We divided the classes into the following two sets. (1) Positive set: Classes in both DBO 3.2 and DBO 2014. (2) Negative set: Classes that are only in DBO 2014. Figure 4 shows the F1 scores with respect to Gold-standard 1 and Gold-standard 2 as binary classifications. Based on these results, the size (ρ) of the multilingual pivotal ontology is set to 260 to obtain the best F1 score performance from the two evaluations. It is possible to reduce the size and hierarchy to only 260 from the top of the final order as a basis for the total of 685 classes. In comparison with Table 2, this smaller ontology may have at least approximately 90% coverage of ABox.

3.2   Result Analysis

In this paper, we focused on constructing a base ontology by reducing the size of a given ontology. The top classes selected through rank aggregation through multilingual analysis were not a top-tier selection on the ontology hierarchy. As shown in Fig. 5, we can see that the overall ontology hierarchy is consistently reduced.

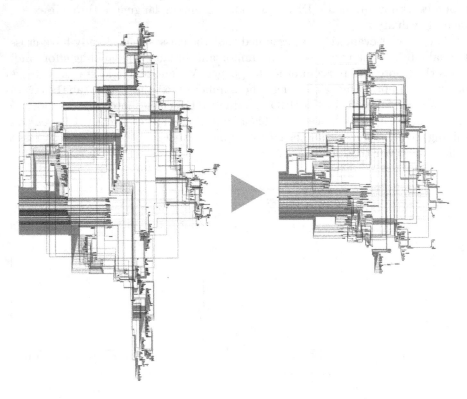

Fig. 5. Comparison of ontology hierarchies between origin and small-sized

4   Conclusion

We presented an approach for identifying global representative classes from DBpedia ontology (DBO), regarded as a multilingual pivotal ontology in this work. We combined the different independently constructed preferences of ranks for each language edition of Wikipedia to produce a consensus order of classes for DBO that is more desirable for representing the knowledge base for multilingual reuse and for connectability as Linked Open Data (LOD) through ontology. Our experimental results showed that the proposed approach significantly helped improve the labeling performance for non-English languages compared to both monolingual and random methods; the selected classes can be smaller than the entire ontology without significant loss of coverage. We expect that a representative subset of DBO in this paper will have a central role in the enrichment and integration of local language knowledge resources in LOD, avoiding islands of monolingual data.

Acknowledgements. This work was supported by Institute for Information & communications Technology Promotion (IITP) grant funded by the Korea government (MSIP) (No. R0101-16-0054, WiseKB: Big data based self-evolving knowledge base and reasoning platform); the Bio & Medical Technology Development Program of the NRF funded by the Korean government, MSIP(2015M3A9A7029735).

References

1. Bizer, C., Heath, T., Berners-Lee, T.: Linked data - the story so far. Int. J. Semant. Web Inf. Syst. **5**(3), 1–22 (2009)
2. Lehmann, J., Isele, R., Jakob, M., Jentzsch, A., Kontokostas, D., Mendes, P.N., Hellmann, S., Morsey, M., van Kleef, P., Auer, S., Bizer, C.: DBpedia - a large-scale, multilingual knowledge base extracted from wikipedia. Semant. Web J. **6**, 167–195 (2014)
3. Kim, E.-K., Weidl, M., Choi, K.-S., Auer, S.: Towards a Korean DBpedia and an approach for complementing the Korean wikipedia based on DBpedia. In: Auer, S., Gray, J., Müller-Birn, C., Pollock, R., Gray, S.W. (eds.) Proceedings of the 5th Open Knowledge Conference, vol. 575, pp. 12–21. CEUR-WS.org (2010)
4. Kontokostas, D., Bratsas, C., Auer, S., Hellmann, S., Antoniou, I., Metakides, G.: Internationalization of linked data: the case of the Greek DBpedia edition. J. Web Semant. **15**, 51–61 (2012)
5. Garcia, J., Montiel-Ponsoda, E., Cimiano, P., Gómez-Pérez, A., Buitelaar, P., McCrae, J.: Challenges for the multilingual web of data. Web Semant. Sci. Serv. Agents World Wide Web **11**, 63–71 (2011)
6. Page, L., Brin, S., Motwani, R., Winograd, T.: The pagerank citation ranking: bringing order to the web. Technical report, Stanford University (1999)
7. Langville, A.N., Meyer, C.D.: Google's PageRank and Beyond: The Science of Search Engine Rankings. Princeton University Press, Princeton (2006)
8. Borda, J.C.: Memoire sur les elections au scrutin (1781)

Identifying Poorly-Defined Concepts in WordNet with Graph Metrics

John P. McCrae[✉] and Narumol Prangnawarat

Insight Centre for Data Analytics, National University of Ireland,
Galway, Republic of Ireland
john@mccr.ae, narumol.prangnawarat@insight-centre.org

Abstract. Princeton WordNet is the most widely-used lexical resource in natural language processing and continues to provide a gold standard model of semantics. However, there are still significant quality issues with the resource and these affect the performance of all NLP systems built on this resource. One major issue is that many nodes are insufficiently defined and new links need to be added to increase performance in NLP. We combine the use of graph-based metrics with measures of ambiguity in order to predict which synsets are difficult for word sense disambiguation, a major NLP task, which is dependent on good lexical information. We show that this method allows use to find poorly defined nodes with a 89.9% precision, which would assist manual annotators to focus on improving the most in-need parts of the WordNet graph.

Keywords: WordNet · Language resource · Data quality · Graph metrics · Lexical resources

1 Introduction

Princeton WordNet [1] is the most widely used lexical resource and even with the recent rise in deep learning and machine learning approaches to NLP, it has been shown [2,3], that the best solutions (such as at SemEval 2016 [4]) to many tasks in natural language processing still rely on this resource. As such WordNet is one of the most vital resources for knowledge extraction and integration. However, there have also been many criticisms of WordNet as an unreliable and error-prone resource and there were significant quality issues ranging from misspellings and cycles in the hypernym graph[1] to issues with poor definitions [5]. Moreover, WordNet is a resource whose principal aim is to use a graph in order to describe the concepts of a language and the methods that build on it normally do not rely on textual descriptions of a concept but only its graph relationships. This is problematic as the average degree of the WordNet graph is only 2.43, which is significantly less than that of similar knowledge graphs such as DBpedia [6],

---

[1] For example: https://lists.princeton.edu/cgi-bin/wa?A2=ind1509&L=wn-users&P=R2&1=wn-users&9=A&I=-3&J=on.

© Springer International Publishing AG 2017
M. van Erp et al. (Eds.): ISWC 2016 Workshops, LNCS 10579, pp. 66–75, 2017.
https://doi.org/10.1007/978-3-319-68723-0_6

which has an average degree of 6.39^2. For some concepts this may be sufficient to describe the meaning of the word, for example the concept 'Slovenian' is described only by the fact that it 'pertains to' the concept 'Slovenia', which in this case is a sufficient description, but for a more complex concept, in particular adverbial concepts such as 'fairly' many links would be required, yet WordNet frequently contains one or even zero links for adverbs.

Princeton WordNet is a manually-developed resource and its value as a gold-standard resource is one of the main reasons that it is so widely used in natural language processing. As such, a fully-automatic approach, such as [7] to the extension or improvement of this resource would not create a resource that can be applied with the reliability of WordNet. Due to recent changes instantiated by the Global WordNet Association to found an interlingual index [8], WordNet is developing from a resource that is developed by one institute for one language into a collaborative project considering multiple language and contributors. As such, it is wise to consider where this collaborative effort is best directed, and this paper's main contribution is to provide a function that can rank every node in the WordNet graph according to whether the description is sufficient for NLP tasks.

In order to estimate the quality of the graph, we use word sense disambiguation (WSD) as a proxy task for representing quality. This is for several reasons, firstly that this has been established by other authors [9] as a suitable task for this purpose. Moreover, it is our intuition that the 'bad' nodes in the WordNet graph are those for which the graph does not provide sufficient information to describe the concept, and thus it follows that a WSD algorithm would also have problem with such concepts. Finally, there have been several methods identified recently [10], which can perform WSD, using only the WordNet graph and without any supervision, while still providing state-of-the-art WSD performance. As such, WSD seems to be the ideal task for the measurement of the quality of individual WordNet nodes.

This work is focussed on WordNet as a particular knowledge graph, as it is the most widely used graph and as it is manually constructed then we have a clear idea of how this analysis can directly help in the lexicon construction process. However, we note that this work is applicable to other forms of knowledge graphs such as DBpedia, and could help in the process of integrating automatically extracted taxonomies, with manually constructed lexicons. Moreover, many of the metrics here generically describe the structure of the graph and could be adapted for semantic similarity or even cross-lingual linking, which is of particular importance for the development of interlingual wordnets.

2   Related Work

The quality of a language resource, such as WordNet, affects its applicability for many tasks and has thus been the focus of many studies. One particular aspect

---

^{2} This is calculated as usual as the number of links (triples) divided by the number of nodes (entities) in the graph.

has been a focus on the technical quality of the resource such as Lohk et al. [11], who looked at the quality of a graph by looking at existing patterns within the graph structure, which may be erroneous, or similar work by Liu et al. [12]. Other work on technical quality has focused on detecting empty, duplicate and logically unsound structures in wordnets [13]. Nadig et al. [14] examined the semantic correctness issue, in particular looking to validate if links in the graph could be validated based on corpus, definition or structural information. Another corpus-based approach to evaluating the quality of a wordnet was followed by Krsteve et al. [15]. These works differ from this paper crucially in that they detect where information is likely incorrect, whereas we focus on where data is absent.

Another aspect of quality has been fitting a second taxonomy, especially that of an upper-level ontology to WordNet, such as the work of Gangemi et al. [16], where WordNet was fitted to the DOLCE ontology, which was said to improve the hierarchy of WordNet. Similar to this Kaplen et al. [17] worked on examining the logical errors in wordnet in particular issues such as multiple inheritance and transitive inference of properties. It has however not been clearly shown that these structural issues impact actual applications, however a significant issue that has been detected is to do with *sense granularity*, that is the distinction between similar meanings. It has been shown [18], that a less fine-grained sense distinction is better for WSD and as such a more coarse-grained sense distinction has been used for the construction of wordnets in other languages [19].

3 Methodology

3.1 Word Sense Disambiguation

In order to learn the quality of a single node in a WordNet graph, we need a proxy task in order to understand the effectiveness of the graph around a given concept. For this we use WSD and in particular we used the Personalized PageRank (PPR) algorithm developed by Agirre et al. [10]. We ran the standard mode of the PPR algorithm for every sentence in the Brown corpus, based on the sense annotation given in SemCor[3]. For each synset in the graph we aggregated the results of the output per synset, that is we counted the precision as how many times out of its occurrences in the gold-standard Brown corpus it was correctly identified by the PPR algorithm. For synsets that did not occur in the Brown corpus, we treated the precision as a missing value and did not use to learn the quality estimator (Table 1).

In Fig. 1, we see the comparison between the frequency of the synsets in the Brown corpus and the precision that was obtained in the WSD task. We observe that there is very little difference in performance for the higher-frequency concepts than for the low-frequency concepts. Instead in Fig. 2, we compare the frequency to the node degree and in this case we see a very different result, suggesting that for the particular method of PPR, the degree of the node is a key

---

[3] http://web.eecs.umich.edu/~mihalcea/downloads.html#semcor.

Table 1. Statistics about Princeton WordNet 3.0 and the Brown corpus

Number of synsets	117,791
Number of links	285,688
Annotations in Brown corpus	234,136
Synsets at least once in Brown	31,755

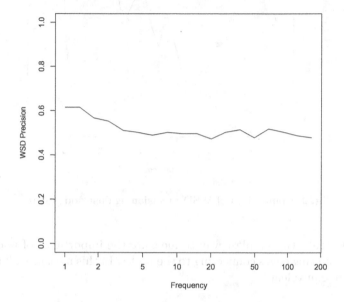

Fig. 1. Comparison of WSD precision against frequency

predictor for the quality of a WordNet node. Both these graphs we generated by taking the precision of the WSD for synsets grouped by their degree or frequency.

3.2   Graph-Based Metrics

Wordnet graph is constructed as a directed typed[4] graph $G = (V, E)$. V is a set of nodes where each node represents a synset s and E is a set of edges where each edge e_{ij} connects synset i and synset j that have any semantic relations. In other words, $e_{ij} \notin E$ if no semantic relation between synset i and synset j.

We observed (Fig. 2) that the higher degree of the Wordnet synset, the better precision of WSD is, however the actual correlation of degree and WSD precision is very low. We would like to combine other graph properties to increase precision of WSD and introduce graph measures that we analyzed as features in this task.

---

[4] The type of the links, such as 'hypernym', are ignored in this work.

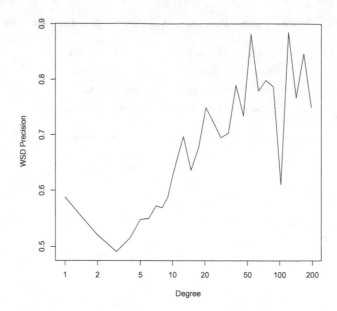

Fig. 2. Comparison of WSD precision against node degree

Degree. Degree is the simplest way to measure the importance of nodes in the network by counting edges connecting to the nodes. In this measure, all neighbors of a node are equivalent.

$$d(s) = |\{(s, s_j) \; in E\}|$$

Network Centralities. One of the most important measures to rank the importance of nodes in a graph are centrality measures. We measured the following network centralities[5]:

Betweenness centrality measures a node by considering the shortest path from a node to itself.

Closeness centrality measures how far is a node to any other node in the network by considering the average distance from the node to every other node in the graph.

Eigenvector centrality [20] measures centrality of a node based on the centrality of its neighbors from the idea that a node becomes more important if it is connected to the important nodes, which can be found from the eigenvector of the adjacency matrix.

PageRank [21] normalizes centrality by dividing the centrality of a node by the number of the nodes it points to and distributing equally to them. The idea is that an important nodes may point to many different nodes but all its

---

[5] We use the implementations provided by NetworkX (https://networkx.github.io) for our analysis.

neighbors are not necessarily considered as high important nodes. A node with high centrality that points to many other nodes will pass a small amount of its centrality to the others.

Average Degree of Neighbors. The average degree of neighbors of synset, where n is the number of the neighbors of synset s, is:

$$avg(d) = \frac{1}{n}\Sigma d(s_i)$$

The higher neighbor degree, the more information we can acquire from the synset neighbors.

Cycles and Triangles. A cycle is a sequence of edges where the first node and the last node are the same node.

In particular, we analyzed the cycles of length 3 which are called triangles.

$$triangle(s) = \{(s, s_1, s_2); (s, s_1) \in E \wedge (s, s_2) \in E \wedge (s_1, s_2) \in E\}$$

Triangles can reveal how many synset neighbors have semantic relations with nearby neighbors whereas cycles can include distant neighbors.

Cluster Coefficient. The cluster coefficient measures the likelihood that the neighbors of each node will connect with each other. This measure is also used to find which nodes tend to be clustered together as relevant synsets.

We analyzed the cluster coefficient of a synset s as the following equation:

$$clust_s = \frac{2 \times triangle(s)}{d(s)(d(s) - 1)}$$

where $triangle(s)$ is a number of triangle of article s and $d(s)$ is the degree of the article s.

Many features, such as degree, exhibit a power-law distribution, therefore we experimented with applying log to all features and took the best performing version of the metric.

3.3   Word-Based Metrics

In addition to graph-based metrics, much of the precision of WSD depends on whether the synset is ambigiuous. To this end, we developed features that decide how ambiguous a particular synset is. The first measure is the log synset size defined for a synset $s = \{w_1, \ldots, w_n\}$

$$\text{log-size}(s) = \log(|s|)$$

Next we define the ambiguity of a word w as the number of synsets that w is part of, e.g.,

$$\text{ambig}_w(w) = |\{s : w \in s\}|$$

We then use log average ambiguity as follows:

$$\text{ambig}_s(s) = \log\left(\frac{\sum_{w_i \in s} \text{ambig}_w(w_i)}{|s|}\right)$$

Let $f(w, s)$ denote the frequency of word w with sense s in the Brown corpus. We denote such a sense as a most frequent sense, mfs, as

$$s \in \text{mfs}(w) \Leftrightarrow f(w, s) = \max_{s'} f(w, s')$$

Finally we define the MFS score for a synset as the percentage of senses for which it is the MFS

$$\text{mfs-score}(s) = \frac{|\{w \in s \wedge s \in \text{mfs}(w)\}|}{|s|}$$

4   Results

In Table 2 we see the correlations between the individual features and the precision as predicted by the word sense disambiguation task, evaluated using 10-fold cross-validation. For these features we see that in general there is low correlation across all the features. This implies that no single measure of quality can be used to estimate whether a node will perform well at predicting precision for the WSD task.

Table 2. Correlation of individual features with precision and log degree

	Precision	Log degree
Log degree	0.081	1.000
Closeness centrality	−0.096	0.463
Log average neighbor degree	−0.166	−0.164
Log number of cycles	−0.006	0.540
Log number of triangles	0.014	0.572
Log Eigenvector centrality	−0.028	0.429
Log PageRank centrality	0.142	0.980
Log betweenness centrality	−0.015	0.770
Log clustering coefficient	−0.012	0.420
Synset size	0.019	0.208
MFS score	−0.526	0.181
Ambig$_s$	0.436	−0.134

Table 3. Prediction of WSD precision based on features

Features	Classifier	Correlation	Absolute error	Mean squared error
Graph features	Linear	0.2341	0.4367	0.4600
Word features	Linear	0.5556	0.3258	0.3934
Both features	Linear	0.6319	0.3018	0.3667
Graph features	Tree	0.3964	0.3910	0.4344
Word features	Tree	0.5725	0.3154	0.3879
Both features	Tree	0.6795	0.2569	0.3472

Table 4. Ranking by WordNet expert of top 50 and bottom 50 synsets

	Top 50	Bottom 50	Average predicted precision
Completely lacking	1	36	0.07
Majorly lacking	0	5	0.05
Slightly lacking	34	8	0.80
Sufficient	14	1	0.92

Following this evaluation we combined all these features using two classifiers: A linear regression model, and the M5P decision tree algorithm [22] (all implementations were those provided by Weka[6]), the results are presented in Table 3. We present the Pearson's correlation (higher is better) as well as both the average absolute error and the mean squared error (lower is better). We also analyzed the effects of just the graph-based features (Sect. 3.2) and the word-based features (Sect. 3.3). We see that the word-based features are more important for predicting precision, however this is unsurprising as these features directly measure the ambiguity of a particular word. The purely graph-based features, however, still show strong performance and for all classifiers the combination of graph-based features and word-based features significantly outperforms other features.

Finally, to evaluate whether this achieves the goal of identifying low and high quality node, an expert on WordNet evaluated the top 50 highest scoring and top 50 lowest scoring entities. This was performed as a double-blind experiment, where the annotator had to rate the entries as "Completely lacking" if there were no semantic relations, "Majorly lacking" if there were only one or two semantic relations, "Slightly lacking" if was either a diverse set of relations or there were many relations of the same type (typically only 'hypernym'/'hyponym' relations) and "Sufficient" if there were many links of different types. The results are presented in Table 4 and show that our system can with high precision detect those nodes in need of improvement. The table also shows the average predicted precision score given by our system for each of the categories indicating a correlation between our systems evaluation and the annotator's opinion.

---

[6] http://www.cs.waikato.ac.nz/ml/weka/.

5   Conclusion

We have presented a system for identifying nodes that have insufficient description in Princeton WordNet. We followed a model where we regressed a number of features to the per-synset precision on WSD. Two sets of features were examined: firstly, graph-based features looking at the structure of the wordnet graph around the node and secondly, word-based features, which measured the ambiguity of the synset. We found that both features were complementary and that the combination of these features was effective at predicting the quality of nodes. Our features do not yet consider the actual type of links in the wordnet graph and as future work, we plan to include these into our evaluation.

References

1. Fellbaum, C.: WordNet. Wiley Online Library, New York (1998)
2. Rothe, S., Schütze, H.: Autoextend: extending word embeddings to embeddings for synsets and lexemes. In: Proceedings of the 53rd Annual Meeting of the Association for Computational Linguistics. Long Papers, vol. 1 (2015)
3. Rychalska, B., Pakulska, K., Chodorowska, K., Walczak, W., Andruszkiewicz, P.: Samsung Poland NLP team at SemEval-2016 task 1: necessity for methods to measure semantic similarity. In: Proceedings of the 10th International Workshop on Semantic Evaluation, pp. 614–620 (2016)
4. Agirre, E., Banea, C., Cer, D., Diab, M., Gonzalez-Agirre, A., Mihalcea, R., Rigau, G., Wiebe, J.: Semeval-2016 task 1: semantic textual similarity, monolingual and cross-lingual evaluation. In: Proceedings of the 10th International Workshop on Semantic Evaluation, pp. 509–523 (2016)
5. Bond, F., Vossen, P., McCrae, J.P., Fellbaum, C.: CILI: the collaborative interlingual index. In: Proceedings of the Global WordNet Conference 2016 (2016)
6. Lehmann, J., Isele, R., Jakob, M., Jentzsch, A., Kontokostas, D., Mendes, P.N., Hellmann, S., Morsey, M., van Kleef, P., Auer, S., et al.: DBpedia-a large-scale, multilingual knowledge base extracted from Wikipedia. Semant. Web 6(2), 167–195 (2015)
7. Navigli, R., Ponzetto, S.P.: Babelnet: the automatic construction, evaluation and application of a wide-coverage multilingual semantic network. Artif. Intell. 193, 217–250 (2012)
8. Vossen, P., Bond, F., McCrae, J.P.: Toward a truly multilingual Global Wordnet Grid. In: Proceedings of the Global WordNet Conference 2016 (2016)
9. Cuadros, M., Rigau, G.: Quality assessment of large scale knowledge resources. In: Proceedings of the 2006 Conference on Empirical Methods in Natural Language Processing (2006)
10. Agirre, E., Soroa, A.: Personalizing Pagerank for word sense disambiguation. In: Proceedings of the 12th Conference of the European Chapter of the Association for Computational Linguistics, pp. 33–41. Association for Computational Linguistics (2009)
11. Lohk, A., Fellbaum, C., Vohandu, L.: Tuning hierarchies in Princeton WordNet. In: Proceedings of the Global WordNet Conference (2016)
12. Liu, Y., Yu, J., Wen, Z., Yu, S.: Two kinds of hypernymy faults in wordnet: the cases of ring and isolator. In: Proceedings of the Second Global WordNet Conference, pp. 347–351 (2004)

13. Smrž, P.: Quality control for wordnet development. In: Proceedings of the Second International WordNet Conference (2004)
14. Nadig, R., Ramanand, J., Bhattacharyya, P.: Automatic evaluation of wordnet synonyms and hypernyms. In: Proceedings of ICON-2008: 6th International Conference on Natural Language Processing, p. 831 (2008)
15. Krstev, C., Pavlović-Lažetić, G., Obradović, I., Vitas, D.: Corpora issues in validation of Serbian WordNet. In: Matoušek, V., Mautner, P. (eds.) TSD 2003. LNCS, vol. 2807, pp. 132–137. Springer, Heidelberg (2003). doi:10.1007/978-3-540-39398-6_19
16. Gangemi, A., Guarino, N., Masolo, C., Oltramari, A.: Sweetening WORDNET with DOLCE. AI Mag. **24**(3), 13 (2003)
17. Kaplan, A.N., Schubert, L.K.: Measuring and improving the quality of world knowledge extracted from wordnet. University of Rochester, Rochester (2001)
18. Yong, C., Foo, S.K.: A case study on inter-annotator agreement for word sense disambiguation. In: Proceedings of the ACL SIGLEX Workshop on Standardizing Lexical Resources (SIGLEX 1999), College Park (1999)
19. Carpuat, M., Ngai, G., Fung, P., Church, K.: Creating a bilingual ontology: a corpus-based approach for aligning WordNet and HowNet. In: Proceedings of the 1st Global WordNet Conference, pp. 284–292 (2002)
20. Bonacich, P.: Power and centrality: a family of measures. Am. J. Sociol. **92**(5), 1170–1182 (1987)
21. Page, L., Brin, S., Motwani, R., Winograd, T.: The PageRank citation ranking: bringing order to the web. Technical report, Stanford InfoLab (1999)
22. Quinlan, J.R., et al.: Learning with continuous classes. In: 5th Australian Joint Conference on Artificial Iintelligence, Singapore, vol. 92, pp. 343–348 (1992)

Extracting Process Graphs from Medical Text Data

An Approach Towards a Systematic Framework to Extract and Mine Medical Sequential Processes Descriptions from Large Text Sources

Andreas Niekler[✉] and Christian Kahmann

Department of Computer Sciences, Natural Language Group, University of Leipzig, Augustusplatz 10, 04109 Leipzig, Germany
{aniekler,kahmann}@informatik.uni-leipzig.de

Abstract. In this paper a natural language processing workflow to extract sequential activities from large collections of medical text documents is developed. A graph-based data structure is introduced to merge extracted sequences which contain similar activities in order to build a global graph on procedures which are described in documents on similar topics or tasks. The method describes an information extraction process which will, in the future, enrich or create knowledge bases for process models or activity sequences for the medical domain.

Keywords: Relation extraction · Natural language processing · Graph processing · Process models

1 Introduction

Medical publications, surgical procedure reports or medical records typically contain procedural descriptions. For example, all activities included in a medical study must be documented for reproducibility purposes, in surgical reports a stepwise description of included procedures is documented and in medical records a history of medical treatment is listed. Additionally, related studies or reports describe alike activities with some alterations or rely on preceding activities that may be described in other documents. This kind of knowledge can be contained in large document collections like the PubMed Dataset.[1] For example, the preparation steps before DNA could be sequenced are often the same but need to be documented for each study. Such redundant activity descriptions can be found amongst many documents describing research within the same domain or field of research. Nevertheless, differences amongst the activities in related documents also exist. A complete overview of activities from a defined document collection provides an easy insight to workflows and paradigms within a domain or field of study. For example, consider the following text snippets extracted from three different documents within the PubMed Dataset.

---

[1] http://www.ncbi.nlm.nih.gov/pubmed/.

© Springer International Publishing AG 2017
M. van Erp et al. (Eds.): ISWC 2016 Workshops, LNCS 10579, pp. 76–88, 2017.
https://doi.org/10.1007/978-3-319-68723-0_7

- aCL and B2GP-I autoantibodies were evaluated at baseline and at 3 and 6 months after the beginning of infliximab treatment. Statistical analysis was performed using Statistica 7.0 PL software. Differences between groups were analyzed using Mann-Whitney U test. A p value less than 0.05 was considered to be statistically significant. We observed 4 aCL IgM-positive (12.5%) patients before the beginning of infliximab treatment.
- The statistical analysis was performed using the SPSS 110 package program. Differences between groups were analyzed using the Mann-Whitney U test. Correlation analyzes were performed using Pearson's correlation test.
- In the event of discordant scores, which differed by a maximum of 1 point, the mean of the two scores was used. Because the semiquantitative data are nonparametric, these data are presented as median (range). Differences between groups were analyzed using the nonparametric Mann–Whitney U-test. For markers scored as present or absent, the χ^2 test was used.

As one can see from the examples, all documents contain the application of the *Mann–Whitney U-test*, even though it is expressed slightly different in the texts. This repeatedly used activity is concurrently used with other activities within the documents. Thus, finding this link between the documents and aligning the activities w.r.t. redundant activities helps to structure and analyze procedural knowledge from topical- or domain-related medical texts. For example, early stages or parts of a larger process might be documented separately to other parts or later stages. In order to extract complete and connected descriptions of such procedural knowledge it is a promising approach to utilize links between different documents and connect the extracted knowledge accordingly.

In this paper a general natural language processing workflow to extract sequential activities from large collections of medical text documents is developed. A graph-based data structure is introduced to merge extracted sequences which contain similar activities in order to build a global graph on procedures which are described in documents on similar topics or tasks. After the review of related work in Sect. 2 the paper introduces the approach in Sect. 3. To demonstrate the potential of such a general workflow we introduce a working example and possible applications on the basis of a subset of the PubMed dataset in Sect. 3.

2 Related Work

The extraction of procedural knowledge from text documents has been investigated for different domains. For example, [2,4,12] describe the process and activity extraction from text as natural language processing (NLP) pipeline. They apply a static rule set on the available features from the results of the NLP pipeline in order to construct the procedural models. In general, the described NLP pipelines use sentence separation, tokenization, POS-tagging and a sentence parser. Other techniques for named-entity-or multi-word-unit-detection are also mandatory for this task. [2,12] apply their methodology to the cooking recipe domain and extract procedural models from single recipe descriptions whereas [4] applies the techniques to different domains with promising results. Additionally, anaphora resolution is also applied in order to match a result of an activity, e.g. the combination of different components like ingredients to prepare a "sauce", to later occurrences of that result in the text which might be references with a different token. Other works try to model processes and activities from tutorial instructions given in natural language or utilize use-case descriptions from requirement specifications [5,13]. In those cases the approach concentrates on a domain and the process description is limited to a fixed set of possible activities.

The creation of probabilitic graphical models using multiple medical records has been investigated in [6]. In this work the authors extract medical problems, tests and treatments from Electronic Medical Records. The extracted information is encoded within a graph structure where the associations between the different types of entities are modeled with co-occurrence statistics. The result of this process is transferred into a probabilistic graphical model which can be used to infer most likely treatments and tests for a medical problem. This work is highly related to the methods described in the paper presented here. However, there are differences in the addressed requirements and properties of the data. The work of [6] builds on the fact that different diseases and their according treatment and testing strategies are contained redundantly in the records. This allows to extract co-occurrence statistics among the mutually used medical concepts in different medical records to determine the strength of their association. The approach is focused and tailored to the domain of medical records and addresses the properties of this text source. The workflow described in our paper yields a general approach to the problem of procedural knowledge extraction for different domains. Thus, the co-occurence information of mutual used activities can be very sparse and the chaining and linking of the extracted entities and concepts are addressed in a different way.

With the exception of [6] all examples create process models from single documents. The combination of knowledge and process descriptions from multiple documents is rarely studied. The proposed method in our paper concentrates on the integration of multiple dependent activity sequences found within a domain or text collection. Thereby, we do not fit our methodology to the properties of a text source or domain and mainly use uninformed approaches. The objective of our methodology is the general extraction of global activity sequences from text sources relating to a domain, work field or task of choice.

3   Text Mining Methodology for Process Extraction

In this section we describe a methodology which extracts and links activities from medical text documents. The described system follows a sequence of procedures in order to create an activity graph as a result. First, the text sources have to be processed in order to access the entity items in the text. Different entities in a sentence are related and form an expressed activity. Therefore, the extraction of valid relations that form activities is introduced to the text processing step. The expression of the activities could vary throughout different text collections. To adopt to such properties we describe an active learning process with a Support Vector Machine classification. This approach supports a semi-automated and fast creation of training examples for the classification task of relations in our proposed methodology. The detected activities within single documents can be represented as vertices in a directed graph. This representation is based on the fact that the data structure must reflect temporal relations among the activities such as sequences. Thus, the second step in our proposed methodology is the creation of a directed graph structure which can be further used for the representation of the activities contained within a text collection. In the following section the text processing and the graph creation are discussed in detail.

3.1   Text Processing and Classification for Activity Extraction

The text sources must be separated into sentences and tokens first by using state of the art tools.[2] Additionally, POS-Tagging was applied to the text sources. To extract the procedural knowledge from the texts, named entity recognition (NER) is required as a pre-processing step. Many NER-algorithms for different purposes have been studied. The state of the art ranges from conditional random field classifiers to ensemble learners which combine multiple entity detection algorithms [3,8]. It would be possible to use 3rd party named-entity detection tools in order to annotate entities automatically but the quality depends on the text source in combination with the algorithm. Since this paper describes a mechanism for using annotated entities to extract activities from text documents it isn't the main focus to vote for a single NER-solution. For simplicity and understandability the experiments in this paper were implemented using a standard pattern-based entity detection to put explanations about the decisions for a specific NER-solution aside. A typical pattern for the detection of entities in the medical domain is (adjective* noun+) which identifies all nouns as entities and, in addition, identifies multi-word-units which consist of a sequence of adjectives followed by a sequence of nouns.[3]

In the separated and preprocessed sentences multiple entities may form an activity. Consider the sentence "Real-time_JJ PCR_NNP was_VBD done_VBN using_VBG the_DT fluorescent-labelled_JJ oligonucleotide_NN probes_NNS". Following the pattern for entity extraction given in the above section the entities "Real-time PCR" and "fluorescent-labelled oligonucleotide probes" are extracted from the sentence. The two entities form the activity "done" which can be part of a chain of activities document throughout multiple documents.

The characteristics of activities or relations between entities change within different domains or described procedures. Thus, the process for identifying and

---

[2] OpenNLP was used to process the text sources for this paper. http://opennlp.apache.org/.

[3] The "*" implies a minimum occurrence of 0 and an unbounded maximum occurrency. The "+" implies a minimum occurrence of 1 and an unbounded maximum occurrence.

connecting entities to activities within the sentences should not be fixed or static. To answer this fact the identification of relations or activities is defined as classification task using a Support Vector Machine (SVM) along with word- and POS-Tag-level features [7]. If a sentence contains an entity E_1 and E_2 the two words before E_1, the two words after E_2 and all words between E_1 and E_2 are extracted as features. Furthermore the POS-tags of the extracted words are used as features for the SVM. To name the features the extracted words are prefixed with a feature name. For example, if one word between E_1 and E_2 is "using_VBG" the word gets the prefix "BETWEEN_" and will become the feature "BETWEEN_using". The same procedure is applied with the POS-Tag of this word to form the feature "BETWEEN_VBG". The feature set for each relation in the training data is joined into an example-feature-matrix in order to train the SVM.

Before the training process is applied the user must define the type and the form of the desired relation. On the basis of this definition training examples are collected from the data. For this purpose an active learning procedure is introduced where the user iteratively collects training data with the support of an automatic classification. An initial search for sentences that include a minimum of entities and verbs that indicate an activity is conducted.[4] The search is implemented using a customizable pattern which may be altered w.r.t. different domains and relation types. The set of matching sentences which contain this custom pattern is presented to the user. Correct entities are selected from the proposed sentences along with the definition whether there is a relation between them or not. The features are extracted automatically and the set of positive and negative examples is used to train an initial SVM model. The trained model is used to identify additional examples in the data. The user judges on those examples and with every batch of new examples the classifier can be refined.[5] If the training quality of the SVM does not change with new examples a final model is trained and applied to all documents. The result is a set of sentences from a document collection where each sentence contains an activity or valid relation between entities.

3.2   Process Graphs for Activity Representation and Processing

In the next processing step a data structure is constructed on the basis of the set of activities that were identified by the classification process. For each activity the two entities E_1 and E_2, the Verb V (past participle between them, a document

---

[4] A basic pattern for this purpose is given by the pseudo-pattern (adjective* noun+) ... (using_VBG) ... (adjective* noun+) where the "..." indicate optional words that may be contained between E_1 and E_2.

[5] To implement an active learning process one must simply present positive and negative classification results to the user. After the judgment, the training set can be refined and extended. A new model can be trained for the next iteration of the active learning procedure and new examples can be classified and presented to a user. The presentation and feedback mechanism can be implemented using a graphical user interface or simple command line interactions.

identifier and a sentence identifier are stored[6]. A graph structure A, a directed graph, is introduced where all identified activities are represented as vertices. All vertices that build a sequence of activities within a document are connected with directed edges, e.g. consecutive activities will be connected as a chain of activities within the graph structure. For example, consider the following sentences.

- Pathological_JJ diagnosis_NN of_IN patients_NNS with_IN atherosclerosis-RNA_NNP extraction_NN from_IN biopsies_NNS was_VBD done_VBN by_IN the_DT Qiagen_NNP Kit_NNP protocol_NN .
- RNA_NNP was_VBD cleaned_VBN from_IN DNA_NNP contamination_NN using_VBG DNAse_NNP Qiagen_NNP .
- Reverse_VB transcription_NN was_VBD done_VBN using_VBG Promega's _NNP reverse_VB transcriptase_NN M-MLV_NNP protocol_NN .
- Real-time_JJ
  PCR_NNP was_VBD done_VBN using_VBG the_DT fluorescent-labelled_JJ oligonucleotide_NN probes_NNS .
- Reaction_NN was_VBD done_VBN using_VBG the_DT chemical_NN supplies_NNS manufactured_VBN by_IN the_DT company_NN Eurogene_NNP .

Those examples can be seen as a sequence of activity descriptions from one document and will be connected as a sequence using directed edges between subsequent relations, e.g. vertices in the graph. This procedure creates a chain of connected vertices for every document in A. The main target for the further processing of A is the linking of different activity chains from multiple documents. This will produce a graph structure which represents networks of activities that supplement each other. In A the connected components can be understood as a summary of activities which come from, or lead to, similar activities. For example, multiple surgical reports contain many redundant descriptions for a certain type of surgical procedure. In some cases there might have been complications and the surgeon had to react on those. Those complications are included in a report between two relations R_a and R_b which might be subsequent in other documents describing the same procedure without complications. A graph which merges different sequential activities from different documents should introduce a direct edge and a cycle of activities between R_a and R_b describing the additional complications. In a later review of the graph this cycle represents single differences from the analyzed standard procedure.

To detect similar relations throughout different documents a similarity operation $sim(R_{D_1}, R_{D_2})$ is defined. This similarity operation can be constructed on the basis of word level similarity or semantic similarity. With a preprocessing of the corpus like word2vec or a co-occurrence analysis each of the relation components can be augmented by semantic vectors representing the associated vocabulary, e.g. the semantic embedding [1, 10]. This allows to compare entities semantically and conceptual similarities between entities can be used to find

---

[6] The verb V could also be seen as the modifier or the name of the activity and could be replaced in other tasks. The usage of V (past participle) works for the examples in this paper and can be different in other domains.

alike relations. Note, that the similarity function is another exchangable component of the information extraction approach described in this paper. It can be altered for differnt sources or domains in order to achieve an optimal quality. For simplicity and to concentrate on the graph processing methodology a Jaccard similarity based on character 3-grams is used as similarity function for the examples in this paper. The similarity between all activities is calculated for E_1, E_2 and V separately which results in three different similarity matrices. In the data, similar entities can consist of multiple words and some additional abbreviations in parenthesis which introduces some slight differences amongst them. The usage of character 3-grams is robust for such little variations. The resulting three similarity matrices are transformed to adjacency matrices by applying a threshold to the values. All similarity values that exceed the threshold will be set to 1, the indicator for an edge between two relations. Values beneath the threshold will be set to 0 to indicate no similarity between two relations. It is also imaginable to set three different thresholds for each single similarity matrix or to weight the matrices for further processing. All resulting adjacency matrices are multiplicated element-wise in order to create a single adjacency matrix S of similar activities, e.g. two activities where E_1, E_2 and V are similar between the two activities are represented by the value of 1 in the final matrix.

In the following step the activities considered to be similar are collapsed using the adjacency matrix S resulting in a graph A'. This process is sketched in Fig. 1. Starting from graph A all edges from similar vertices are taken over to a single vertex and all vertices where the edges where taken from are deleted. That means similar vertices are collapsed to a single vertice and the associated ingoing

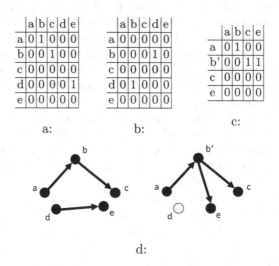

Fig. 1. Collapsing and merging vertices based on similarity information in S with (a:) an Example of the adjeceny matrix of A for document based activity sequences, (b:) adjacency matrix S between similar relations and (c:) the resulting matrix A' after the collapsing of A. In (d:) the merging and edge transfer between two vertices is displayed.

and outgoing edges of those relations are merged. The resulting graph connects the sequences of different documents where similar activities build single vertices with more than one incoming or outgoing edge ($d_G^+(v) > 1$ or $d_G^-(v) > 1$). Activities with this property are identified more frequently than other activities in the data and thus are of some importance for the overall activity summarization. In summary, it can be said, A' is an unconnected graph where a set N of connected components can be identified. This set represents different graphs where the interaction and coherence of related processes, described in different documents, is encoded.

4   Applications and Examples

The resulting graph can be exploited for different applications. In the following examples three possible applications of exploratory data analysis are discussed. All examples are created on the basis of data from the PubMed dataset which was additionally reduced to a subcorpus consisting of 2.813 documents. The documents all contain the keyword phrase "Ankylosing spondylitis"[7], which represents an autoimmune disease of the axial skeleton.

4.1   Summarization of Activities as Process Graphs

In the first example the method is used to extract sequences of activities from studies in a specific domain. The corpus was separated into sentences and word-tokens. Additionally, POS-tagging was applied with the PENN Tagset.[8] The processing is started by looking for sentences containing the pattern (adjective* noun+) ... (was,were,has,been,had)_VBD) ... (using_VBG) ... (adjective* noun+). This pattern can be seen as a user defined constraint which could be altered for other text sources or domains. In this case the pattern reflects stereotype sentences from the corpus which describe an activity that has been carried out by the authors of the underlying medical paper. The user decides which of the matching sentences suit the defined or required description of an activity. All validated examples are passed to an initial training set for the SVM classifier described in Sect. 3.1. After this step an active learning process is applied and the training set is extended semi-automatically. The overall process, including all generated training examples, identifies 14.087 relations from the corpus which will be further processed. The next step links similar relations from the classification result as described in Sect. 3.2. For this example the threshold for the Jaccard similarity of R_{D_1} and R_{D_2} is set to 0.5. Afterwards, the graph adjacency matrix of A is created and the edges for the inner-document connections of the relations are inserted. The similarity matrices for all E_1, E_2 and V are multiplied element-wise to find similar relations and A is collapsed by the similarity information to produce the matrix A'. The resulting adjacency matrix is converted

---

[7] http://en.wikipedia.org/wiki/Ankylosing_spondylitis.
[8] http://www.comp.leeds.ac.uk/ccalas/tagsets/upenn.html.

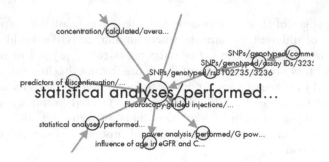

Fig. 2. A graph visualization centered on the central activity "statistical analyses/performed/software".

to a graph. The graph contains a set of connected components. Each connected component can be seen as a single graph which represents a separated summary of activities. The single connected components can be visualized and explored separately by an analyst. To filter for activity summaries containing prominent relations the components are only kept if they contain at least one vertice which is based on a relation that was found more than 3 times in the data. Furthermore all components which do not contain relations from documents with a given set of keywords included are filtered out. This additional procedure allows to drill down the analysis to a user defined focus. For this example the keywords "gene" and "tissue" where used to filter out graph components drawn from documents not containing those words. The initial graph A of the given example consists of 14.987 vertices and 23.902 edges. The graph contains 1385 connected components with a minimum of 1 edge. The median diameter among all connected components is 6. The final graph A' is reduced to 10.453 vertices and 9.234 edges. This processed version of the graph contains 1.063 connected components. The median diameter among all connected components in A' is 10. As one can see, the diameter of the connected components rises and the procedural knowledge among different documents is linked within the final graph.

The resulting graphs can be visualized and analyzed. In this experiment Gephi is used for visualization purposes [9]. Within Gephi a graph could be further processed and explored. For example, a user can filter the graph for vertices that have a certain degree on incoming and outgoing edges. The final visualizations are very useful to summarize and understand the activities which are normally hidden within large document collections. In Fig. 2 an example of a visualized graph structure is given. It can be seen that different activities can produce data which is undergoing a statistical analysis.

4.2 Summarization of Activities as Lists

On the foundation of the graph A' a summarization of the activities as sorted lists can be extracted. The basic problem in producing a global summary of the activities found in the documents is that fact, that their global position in the whole process is unknown. The only known fact is the relative position w.r.t. direct neighbors in the graph. Those neighbors are normally the preceding or following activities from one document. The graph structure can be used to correct or set a global positioning index for each activity in the following way.

1. First, a sequence of all shortest paths (SP) within a single connected component is built. This process is repeated for all connected components in A'.
2. For each set of SP's an iteration from the longest to the shortest SP is conducted.
3. For one SP the process follows the direction of the edges, starting from the global positioning index of the first vertex, which might be the sentence number from the source document of the underlying relation. All subsequent vertices are forced to have a larger position index than their preceding vertex in the current SP.

Some SP's are overlayed and contain identical vertices. Thus, a vertex can also be included in other SP's. The redundant correction of vertices which are contained in different SP's would lead to a violation of the ascending positioning within a path. Therefore, all possible corrected positions for such a redundant vertex are stored. Remember, that those redundant vertices come from activities which were found several times in the data. The activities finally can be sorted by their global position. The position for activities with multiple position values is averaged. In Table 1 a possible result is sketched. Such a view allows to review different phases of activities in complex processes which were reported within a document collection.

4.3   Information Retrieval Within Process Graphs

The graph structure is also useful for querying information. To query the connected graph components all vertices containing a given keyword are accessed and preceding and following vertices are extracted. The query for "gene" results in a set of vertices containing the information given in Table 2. Of course it is imaginable to select preceding or following vertices which are more than one vertex away from the matching activities. This application allows for the detailed review of activities linked to a user defined concept. The given concept could be a certain technology or method. In turn, the graph could be mined for an activity like "tissue cells" and the prequisite methods and technologies and products of the activity are observable.

4.4   Future Work

This paper describes an idea of an information extraction process which creates global activity descriptions from many text documents. A relation-extraction and relation-connecting workflow based on text mining methods is presented and the experiments show promising results for practical applications which need to be optimized and evaluated in quality and accuracy. The potential of

Table 1. A graph corrected sequence of activities. Activities found several times in the data are printed in bold and their positions are averaged.

Name of activity	Avg. position
Significant differences/determined/analysis of variance	464,5
MNCs (nuclei/counted/light microscopy	464,5
Statistical analyses/performed/software R	467,7
Results/tested/Wilcoxon test	470
P values/presented/Altman and Bland	471,3
Inter-group differences/evaluated/Mann-Whitney U test	474,5
Genesets/identified/test	475
Laser Capture Microdissection/carried/ Zeiss/PALM Microbeam Instrument	477
Figures/plotted/PoseView	479,5
Protein concentration/analyzed/Bradford assay reagent	479,5
Statistical analysis/performed/Prism	480
Statistical analyses/performed/SPSS V	480
Analyses/performed/Stata	480,2
Analyses/performed/Statistical Package for Social S...	480,3
Statistic analyses/conducted/SPSS	481
Analyses/done/SAS software	482,7
Analyses/performed/STATA	484
Statistical analyses/performed/SAS	485,5

user refined learning classifiers for relation classification is highlighted in order to be domain and text source independent. For the merging of activities a very simple similarity function is used for this paper and the accuracy is not optimal. Nevertheless, it is possible to show the potential for useful applications based on the described information extraction process for relations.

In order to quantitatively judge on the quality of the extraction process an evaluation dataset and evaluation strategy needs to be developed as prequisite for future work. More research on suitable similarity functions for relations which can also handle semantic similarities will optimize the quality of the graph merging process. Future work will also include the adoption of domain knowledge from knowledge bases. Those has been described as very helpful resources in order to adopt to a domain in [11]. The links and dependencies between entities and their possible representations in the data can be encoded in those data structures by domain experts. This will add supervision and control to the graph creation process and thus allows for a higher precision of the graph. Additionally, anaphora resolution can be modeled with knowledge bases to connect graph structures where the relations represent processes which produce other entities as results. Such edges can't be established with character or semantic comparison of the relations. In the moment a connection can only be established if the producing process is encoded within a single document. Furthermore, the introduction of manual corrections steps to refine the graph and the improvement of the quality and the transferability of the relation extraction classification may also be promising elements to optimize the quality.

Table 2. Example result for activities including the word "gene". Additionally the incoming and outgoing activities are shown.

Incoming
Data/genotyped/different platforms
Concentration of genomic DNA/measured/ng/
Supernatants/analyzed/eBioscience
Quality controlGenomic DNA/extracted/Puregene DNA Isolation Kit(Gentra Systems , Minneapolis , MN , USA)
PBMCs/counted/CASY cell counter (Roche)
Hit
Major histocompatibility complex region/genotyped/Illumina Infinium 15 K array
Samples/genotyped/ImmunoChi
Healthy donors/genotyped/Applied Biosystems TaqMan SNP
Individual/genotyped/Affymetrix Genome-Wide Human SNP Array
Controls/genotyped/Illumina HumanCNV370-duo chip
Outgoing
Genotype calls/made/BRNNP algorithm
Power calculations/carried/Genetic Power Calculator
Analysis of intensity clusters and genotype calls/performed/Illumina Genome Studio software
Fine mapping linkage study, allele frequencies/estimated/MENDEL software
RNA levels/quantified/Illumina HT-12 V3.0 platform

Acknowledgement. ExB Labs GmbH kindly helped to compile and preprocess the corpus for this paper.

Funding: The project is funded by European Regional Development Fund (ERDF/EFRE) and the European Social Fund (ESF).

References

1. Bordag, S.: A comparison of co-occurrence and similarity measures as simulations of context. In: Gelbukh, A. (ed.) CICLing 2008. LNCS, vol. 4919, pp. 52–63. Springer, Heidelberg (2008). doi:10.1007/978-3-540-78135-6_5
2. Dufour-Lussier, V., Le Ber, F., Lieber, J., Nauer, E.: Automatic case acquisition from texts for process-oriented case-based reasoning. Inf. Syst. **40**, 153–167 (2014)
3. Finkel, J.R., Grenager, T., Manning, C.: Incorporating non-local information into information extraction systems by gibbs sampling. In: Proceedings of the 43rd Annual Meeting on Association for Computational Linguistics, pp. 363–370. Association for Computational Linguistics (2005)

4. Friedrich, F., Mendling, J., Puhlmann, F.: Process model generation from natural language text. In: Mouratidis, H., Rolland, C. (eds.) CAiSE 2011. LNCS, vol. 6741, pp. 482–496. Springer, Heidelberg (2011). doi:10.1007/978-3-642-21640-4_36
5. Gil, Y., Ratnakar, V., Frtiz, C.: TellMe: learning procedures from tutorial instruction. p. 227. ACM Press (2011)
6. Goodwin, T., Harabagiu, S.: Clinical data-driven probabilistic graph processing. In: Proceedings of the Ninth International Conference on Language Resources and Evaluation (LREC 2014), Reykjavik, Iceland, European Language Resources Association (ELRA), May 2014
7. GuoDong, Z., Jian, S., Jie, Z., Min, Z.: Exploring various knowledge in relation extraction. In: Proceedings of the 43rd Annual Meeting on Association for Computational Linguistics, pp. 427–434 (2005)
8. Hänig, C., Bordag, S., Thomas, S.: Modular classifier ensemble architecture for named entity recognition on low resource systems. In: Proceedings of the KONVENS GermEval Shared Task on Named Entity Recognition, pp. 113–116, Hildesheim, Germany (2014)
9. Bastian, M., Heymann, S., Jacomy, M.: Gephi: an open source software for exploring and manipulating networks (2009)
10. Mikolov, T., Chen, K., Corrado, G., Dean, J.: Efficient estimation of word representations in vector space. CoRR, abs/1301.3781 (2013)
11. Roberts, K., Harabagiu, S.M.: A flexible framework for deriving assertions from electronic medical records. J. Am. Med. Inf. Assoc. 18(5), 568–573 (2011)
12. Schumacher, P., Minor, M., Walter, K., Bergmann, R.: Extraction of procedural knowledge from the web: a comparison of two workflow extraction approaches. p. 739. ACM Press (2012)
13. Yue, T., Briand, L.C., Labiche, Y.: An automated approach to transform use cases into activity diagrams. In: Kühne, T., Selic, B., Gervais, M.-P., Terrier, F. (eds.) ECMFA 2010. LNCS, vol. 6138, pp. 337–353. Springer, Heidelberg (2010). doi:10.1007/978-3-642-13595-8_26

Chainable and Extendable Knowledge Integration Web Services

Felix Sasaki[1]([✉]), Milan Dojchinovski[2,3], and Jan Nehring[1]

[1] Language Technology Lab, DFKI GmbH, Berlin, Germany
{felix.sasaki,jan.nehring}@dfki.de
[2] Knowledge Integration and Language Technologies (KILT/AKSW), InfAI,
Leipzig University, Leipzig, Germany
dojchinovski@informatik.uni-leipzig.de
[3] Web Intelligence Research Group, FIT, Czech Technical University in Prague,
Prague, Czech Republic
milan.dojchinovski@fit.cvut.cz

Abstract. This paper introduces the current state of the FREME framework. The paper puts FREME into the context of linguistic linked data and related approaches of multilingual and semantic processing. In addition, we focus on two specific aspects of FREME: the FREME NER e-Service, and chaining of FREME e-Services. We believe that the flexible and distributed combination of e-Services bears a potential for their mutual improvement. The FREME framework is an open source software available for free download (https://github.com/freme-project/).

Keywords: Linguistic linked data · NIF · NLP · Named entity recognition · Semantic enrichment

1 Introduction

This paper presents the current state of FREME, a framework for multilingual and semantic enrichment of digital content. A detailed, general overview of the goals of FREME has been given in [9]. Here we focus on two aspects of FREME: the FREME NER service and chaining of FREME services.

FREME is developed in the EU funded FREME project[1], which started in February 2015 and lasts for two years. The project has two aspects: the development of the FREME framework, transferring technology outcomes from several language and data related projects; and the following four business cases:

1. Authoring and publishing multilingual and semantically enriched eBooks;
2. Integrating semantic enrichment into multilingual content in translation and localisation;
3. Enhancing the cross-language sharing and access to open agricultural and food data; and
4. FREME-empowered personalised content recommendations.

---

[1] See http://www.freme-project.eu.

© Springer International Publishing AG 2017
M. van Erp et al. (Eds.): ISWC 2016 Workshops, LNCS 10579, pp. 89–101, 2017.
https://doi.org/10.1007/978-3-319-68723-0_8

The paper is structured as follows. Section 2 puts FREME into the context of the KEKI workshop. Section 3 provides a general overview of the FREME architecture. Section 4 elaborates on the FREME NER service. Section 5 discusses how different e-Services can be chained together. Finally, Sect. 6 concludes the paper.

2   FREME in Context

The development of the FREME framework can be described (a) in the context of linguistic linked data, and (b) with regards to challenges that arise from the four business cases.

Data related to linguistic and natural language processing. In the paradigm of linguistic linked data, more and more language resources are being published as part of the linguistic linked open data cloud[2]. FREME allows processing data available in the cloud as part of content enrichment workflows, for example to adapt named entity recognition with domain specific data sets.

Linguistic and NLP Ontologies. The LLOD cloud gathers language resources that are represented with standard formats. FREME enrichment workflows make use of the following formats:

- The Natural Language Processing Interchange Format (NIF) [6] to represent data and enrichment information;
- The Internationalization Tag Set (ITS) 2.0[3] to represent metadata for improvement of enrichment workflows; and
- The OntoLex Lemon model[4] to represent lexica, including their meaning with respect to ontologies.

Linguistic linked open data workflows. The LLOD technology stack allows creating NLP and data services in a distributed and decentralized manner. FREME implements this stack by making use of the previously described standards, and by adding a declarative approach to define and re-use enrichment workflows.

NLP techniques for knowledge extraction. One aim of LLOD is to provide techniques for knowledge extraction that deploy linked data. FREME implements this approach in its FREME-NER service and allows users adapting the service with custom datasets, again to be provided as linked data.

Approaches using mappings and their maintenance from semistructured sources. Industry applications of NLP and data enrichment workflows have to deal with a plethora of content formats. Semistructured formats like HTML or certain XML formats are widely used in applications. Via its e-Internationalization service, FREME allows processing these formats, not only

---

[2] See http://linguistic-lod.org/llod-cloud for a latest version of the LLOD cloud.
[3] See https://www.w3.org/TR/its20/.
[4] See https://www.w3.org/2016/05/ontolex/.

for extraction, but for round-tripping, that is: storage of enrichment information in the original format.

The LLOD context of FREME can also be described from the point of view of the four business cases. From the business case perspective, several challenges arise when creating NLP and data processing applications. They are addressed in the following manner by FREME.

Interoperability and chainability. Applications often are provided as silo solutions. Integration of new functionality is then a time consuming task with high integration costs. By using the previously described, standardized technology stack, this effort is reduced significantly. Details are described in Sect. 5.

Adaptability. There is a growing set of applications for key NLP tasks like named entity recognition, see e.g. [7]. Many of them rely on the DBpedia dataset [1] for entity linking. Tools like Stanford NER [5] allow users loading their own dataset and prepare it for NER. However, for users without a technological background in NLP, it is very hard to adapt these tools. FREME eases the adaptation process in several ways, with regards to the configuration of enrichment workflows, usage of custom data sets, and tailoring NER processing towards domains. Details for this adaptation are described in Sect. 4.

Data formats. The four business cases require enrichment workflows in many formats. For example, in localization, the XML based XLIFF format[5] is widely used. Current multilingual and semantic applications allow extraction of content from such formats. However, for real-life applications, the enrichment information has to be stored inside the format, without breaking existing processing tasks like validation, query or transformation. Via the e-Internationalization service, FREME allows such round-tripping processing.

2.1   Related Work

In this section we compare FREME to several related approaches: Apache Stanbol, Weblicht, Apache UIMA, and LAAPS Grid. They offer related capabilities and a comparison helps to understand the role of FREME.

Apache Stanbol offers a set of text analytics services in a Software as a Service manner. It is provided as an open source platform and intends to extend traditional content management systems with semantic services. In addition, the text analytics services can be used within arbitrary applications [2].

Apache Stanbol differs from FREME with regards to the set of services being offered. Although being open source and therefore being theoretically extensible, Apache Stanbol offers no detailed documentation on how to extend it. Further, Apache Stanbol puts a focus on the use case of a semantic content management system and semantic enrichment of homepages. Multilingual enrichment of other types of content is not taken into account. Further, Apache Stanbol offers a variety of low level services like tokenization, part-of-speech tagging and others, so the

---

[5] See http://docs.oasis-open.org/xliff/xliff-core/v2.0/xliff-core-v2.0.html.

enrichment can be performed on different levels of granularity whereas FREME hides theses low level technologies to reduce the complexity of using the services.

Like FREME, *Weblicht*[6] offers support for chainable NLP services in a RESTful manner. The main difference is that Weblicht does not constitute service chains via a linguistic linked data approach. This has the disadvantage that integration with linked data sources into Weblicht services needs an additional software integration step. In FREME no additional software integration is needed, since via the e-Link service, standard linked data query technology (SPARQL) can be deployed. Nevertheless, a conversion between the Weblicht native, XML based TCL format and the linguistic linked data format used in FREME should be possible and has the potential to grow the number of decentralized NLP services.

APACHE UIMA[7] offers a framework for knowledge extraction pipelines. Like FREME, UIMA is extensible with various NLP components. A key difference to FREME again is that UIMA does not provide a linguistic linked data workflow for content enrichment. Instead, like Weblicht, UIMA provides an XML format. Another difference is that UIMA does not come with a Web service layer. This means that access to UIMA is specific to given programming languages (esp. Java or C++). In contrast, FREME can be accessed with nearly all programming languages, since the programming languages only have to offer HTTP request functionality. Since UIMA has a lot of existing modules, like in the case of Weblicht, a conversion between the UIMA and the NIF format could be of great value to benefit from existing NLP services. Further APACHE UIMA offers a variety of low level services like tokenization, part-of-speech tagging and others, so the enrichment can be performed on different levels of granularity whereas FREME hides theses low level technologies to reduce the complexity of using the services.

LAAPS GRID[8] is a framework that enables discovery, composition, and reuse of NLP components. LAPPS GRID comes with certain standards to support NLP tool interoperability. The LAPPS Interface Format (LIF) plays the role of NIF, that is, LIF constitues input and output of NLP workflows. The LAPPS Web Service Exchange Vocabulary defines the terms used in LIF, e.g. for part-of-speech or other layers of linguistic annotation.

LIF is defined as a JSON format, which is a difference to the linguistic linked data approach taken by FREME. In addition, LIF and the terms defined by the vocabulary aim at fostering interoperability of the NLP detailed level processing, e.g. parts-of-speech tagging, tokenization, etc. In FREME, this detailed level is not represented, but rather the higher level output of NLP processes, e.g.: annotated entities, translations, terms, etc. This eases the integration with an application layer and integration with non-linguistic information, provided by the linked data cloud.

The tools and initiatives discussed so far in this section all provide digital content processing functionality. META-SHARE[9] is a distributed network of language

---

[6] http://weblicht.sfs.uni-tuebingen.de/.

[7] http://incubator.apache.org/uima.

[8] http://www.lappsgrid.org/.

[9] http://www.meta-share.eu/.

resource repositories. In the future, the FREME framework itself and resources generated via the FREME project will be made accessible via META-SHARE.

3   FREME Architecture

FREME uses a client-server Web service architecture that exposes Web services, called *e-Services*, via HTTP APIs. This approach allows for a decentralized, distributed creation of services in a RESTful architecture [4]. In this way a combination of services can be flexibly configured instead of being hard-wired in a source code. Further, the technology is not bound to a specific programming language, since almost every programing language supports HTTP based interactions [8]. Additionally, it is designed in an extensible manner, so that any interested party can plug-in more services.

FREME uses common formats for language and data processing workflows, so that e-Services can easily be created by following the linguistic linked data technology stack. In this stack, the NLP Interchange Format (NIF) serves as a common broker format. Both the actual textual content and information generated via NLP and Linked Data processes is stored in NIF.

FREME offers six e-Services. Their functionality is summarized below.

- e-Entity offers named entity recognition. It is discussed in detail in Sect. 4.
- e-Translation offers cloud based machine translation.
- e-Terminology offers enrichment of content with information about terms.
- e-Link offers enrichment with information from the linked data cloud.
- e-Publishing allows storing enriched content in the standardised ePub format.
- e-Internationalisation allows enrichment covering a wide range of digital content formats like HTML, generic XML or selected XML vocabularies.

Each e-Service is a pipeline on its own. For example, e-Entity consists of a series of tasks like word tokenization, part of speech tagging, sentence splitting and more. All these internal steps are hidden from the user. The user just submits text to the service and retrieves the entities. This lowers the complexity to use the service a lot. In some circumstances this might have a negative influence on the processing speed: When several e-Services are executed one after the other, some internal pipeline steps might be repeated.

In addition, the FREME framework is deployed in the German project "Digitial Curation Technologies" (DKT)[10]. Services offered by DKT also use the linguistic linked data technology stack and hence can be combined with FREME services out of the box.

4   Content Enrichment with Names Entities

4.1   FREME NER Overview

E-Entity is one of the most exploited service within the FREME framework. Knowing what entities are mentioned in a document is of essential importance

---

[10] See http://digitale-kuratierung.de/ for details on the project.

to better understand the aboutness of the document. The e-entity service anno-
tates an input document with annotations representing entities. Mentions of
entities, such as people, organizations or locations, are *spotted* and encoded
with their position in the input document. Next, the entity is *disambiguated*
with a type from a set of entity types[11] and linking it to a specified knowledge
base. The spotting and classification step is done by employing the Stanford-
NER tool[12][5] with a trained models on content from Wikipedia. The linking
of entities ultimately relies on the most-frequent-sense approach and links with
the most-frequent-sense entity. FREME NER is currently using models trained
for English, German, Dutch, Spanish, Italian, French and Russian. To realize a
MFS based linking we used Wikipedia as a reference knowledge base and collected every entity surface form, the corresponding hyperlink and the number
of occurrences. As a result, a pair-count dataset [3] which provides this information was generated. The linking step is implemented in Apache Solr[13]. Solr
contains indexed entities with their corresponding URI identifier, possible surface form variations, language, and the dataset they refer to. When performing
the linking step of an entity mention, entity candidates are retrieved according
to their surface form similarity, and the one with the highest `pair count` value
is considered as the correct entity.

Listing 1.1 provides an example of the output from FREME NER.

```
1   <http://freme-project.eu/#offset_0_33>
2              a                nif:Context , nif:OffsetBasedString ;
3          nif:beginIndex  "0"^^xsd:nonNegativeInteger ;
4          nif:endIndex    "33"^^xsd:nonNegativeInteger ;
5          nif:isString    "Diego Maradona is from Argentina."^^xsd:string .
6
7   <http://freme-project.eu/#offset_0_14>
8              a                nif:OffsetBasedString , nif:Phrase ;
9          nif:anchorOf       "Diego Maradona"^^xsd:string ;
10         nif:beginIndex     "0"^^xsd:nonNegativeInteger ;
11         nif:endIndex       "14"^^xsd:nonNegativeInteger ;
12         nif:referenceContext  <http://freme-project.eu/#offset_0_33> ;
13         itsrdf:taClassRef     <http://dbpedia.org/ontology/SportsManager> , <http://
               dbpedia.org/ontology/Person> ;
14         itsrdf:taConfidence   "0.9869992701528016"^^xsd:double ;
15         itsrdf:taIdentRef     <http://dbpedia.org/resource/Diego_Maradona> .
16
17  <http://freme-project.eu/#offset_23_32>
18             a                nif:OffsetBasedString , nif:Phrase ;
19         nif:anchorOf       "Argentina"^^xsd:string ;
20         nif:beginIndex     "23"^^xsd:nonNegativeInteger ;
21         nif:endIndex       "32"^^xsd:nonNegativeInteger ;
22         nif:referenceContext  <http://freme-project.eu/#offset_0_33> ;
23         itsrdf:taClassRef     <http://dbpedia.org/ontology/Place> , <http://dbpedia.org/
               ontology/Location> ;
24         itsrdf:taConfidence   "0.9804963628413852"^^xsd:double ;
25         itsrdf:taIdentRef     <http://dbpedia.org/resource/Argentina> .
```

Listing 1.1. Output from FREME NER in the NIF format.

---

[11] Currently, FREME classifies the entities with four types: PER, ORG, LOC and
MISC for anything else.
[12] http://nlp.stanford.edu/software/CRF-NER.shtml.
[13] http://lucene.apache.org/solr/.

4.2   Entity Linking with Custom Datasets

In the last decade, entity linking has been primarily evaluated on datasets such as DBpedia, YAGO[14] and BabelNet[15]. In these use cases, the entity linking approaches have been exclusively customized to these datasets, and adoption of other datasets requires significant amount of effort, or it is not possible at all. In FREME, we allow users to use their custom proprietary and public datasets and adopt the processing according to their needs.

FREME NER provides a dataset management endpoint which can be used to perform the usual dataset operations such as creation, update and deletion of a dataset. The minimum requirement is to provide a list of entities with a corresponding name variations. This information should be provided in RDF, where the subject of a triple is a URI, which uniquely identifies the entity, and the object is the entity name variation. The name variations can be provided using the RDFS[16] property `rdfs:label` or the SKOS[17] properties `skos:prefLabel` or `skos:altLabel`. While `rdfs:label` and `skos:prefLabel` specify the human-readable version of the entity name, `skos:altLabel` can provide alternative lexical labels for the entities. For example, a `pref:label` for the footballer Maradona is "Diego Armando Maradona", while `skos:altLabel` will be "Maradona".

Note that the confidentiality of proprietary datasets is ensured by implementing a secured access management. A user needs to be authenticated and authorized to get access to a dataset. Thus, only the owners of the particular datasets can consume them.

Currently, FREME NER maintains "general" frequency counts information computed from the DBpedia Abstracts dataset [3]. These frequency counts information is used for the implementation of the most-frequent-sense based entity linking. In our future work, we plan to collect per-domain frequency counts and increase the performance of domain specific NER.

4.3   Domain Specific NER

In long texts, the list of recognized entities can be very large containing also entities which are not relevant to the domain of the document. For example, very often HTML content contains also advertisements in form of text snippets which occur inline with the main content, or it contains entity mentions which are irrelevant for the main content. For example, an HTML document providing recent information about the Syrian crisis might encompass an advertisement related to the UEFA Euro 2016 championship, which mentions a football team or a football player. FREME enables users to filter out such irrelevant entities by specifying the domain of interest (i.e. politics and administration) Thus, only entities from this specific domain will be returned. The implementation of

---

[14] http://www.mpi-inf.mpg.de/departments/databases-and-information-systems/research/yago-naga/yago/.

[15] http://babelnet.org/.

[16] https://www.w3.org/TR/rdf-schema/.

[17] https://www.w3.org/TR/skos-reference/.

this feature is realized by populating list of domains with corresponding entity types[18]. E.g. the types `dbo:PoliticalConcept` and `dbo:PublicService` belong to the domain of politics and administration.

4.4   Experiments

In order to evaluate the (1) *quality* of the enrichments and the (2) *scalability* of the named entity recognition, we have conducted several experiments using GERBIL[19][11], a framework for evaluation of entity annotation tools. The experiments were executed using GERBIL version 1.2.3-SNAPSHOT via the live running instance at http://gerbil.aksw.org/gerbil/. The entity recognition was evaluated on five English collections and one German collection. The collections differ with regards to length of the documents, the density of entity mentions and the topic of the documents[20]. The evaluation was executed without performing domain specific NER. Two types of experiments were conducted in the evaluation. A strong annotation match which requires exact match of the entity mention with the gold standard, and a weak annotation match which requires overlap of the entity mention with the annotation in the gold standard. Table 1 provides detailed results from the experiments for FREME NER. In Table 1 we report the macro and micro measures. The macro measures refer to the performance across the whole dataset, while the micro measures are computed for each documented and then averaged.

The results show that quality of the enrichments depends on the content. The best performance were achieved for the MSNBC dataset with 0.914 F1 for the weak annotation match and 0.805 F1 for strong annotation match. The worst

Table 1. Detailed evaluation results of FREME NER.

Dataset	Lang.	Exp. type	Micro F1	Micro P	Micro R	Macro F1	Macro P	Macro R	Millis per doc	Entities per doc	Millis per entity
Spotlight	EN	Weak	0.349	0.750	0.227	0.278	0.498	0.216	47.83	5.69	8.41
		Strong	0.242	0.520	0.158	0.193	0.317	0.154	35.43	5.69	6.23
KORE50	EN	Weak	0.956	0.940	0.972	0.957	0.958	0.975	31.98	2.86	11.18
		Strong	0.894	0.879	0.910	0.890	0.888	0.909	30.52	2.86	10.67
Reuters-128	EN	Weak	0.813	0.721	0.931	0.808	0.744	0.939	58.84	4.85	12.13
		Strong	0.675	0.598	0.774	0.669	0.614	0.778	53.44	4.85	11.02
RSS-500	EN	Weak	0.677	0.520	0.969	0.736	0.639	0.969	39.48	0.99	39.88
		Strong	0.579	0.446	0.827	0.634	0.552	0.827	34.31	0.99	34.66
MSNBC	EN	Weak	0.914	0.865	0.968	0.893	0.842	0.963	188.55	32.50	5.80
		Strong	0.805	0.763	0.853	0.780	0.738	0.837	164.45	32.50	5.06
News-100	DE	Weak	0.644	0.777	0.550	0.587	0.631	0.571	369.73	22.33	16.56
		Strong	0.447	0.535	0.384	0.373	0.398	0.365	232.42	14.04	16.55

---

[18] See the list of domains and related entity types at https://github.com/freme-project/freme-ner/blob/master/src/main/resources/domains.csv.
[19] http://aksw.org/Projects/GERBIL.html.
[20] More information on the datasets is provided by [10].

performance has been achieved for the DBpedia Spotlight dataset with 0.349 F1 for the weak annotation match and 0.242 F1 for the strong annotation match.

In the experiments we have also evaluated the scalability of the entity recognition, and the evaluation results show that FREME NER in average can process one entity in 15 ms or, in other words, 67 entities per second. Note that this conclusion should be taken with some reserve, since we implement caching. Hence, documents with frequently occurring entities will be processed faster. In [10] the authors report on the time needed to process a document for the MSNBC dataset. According to the results, except for the TagMe 2 system, for all the other systems it took more then one second to process a MSNBC document. In comparison, FREME NER required 914 and 805 milliseconds for the weak and strong annotation match, respectively.

In Table 2 we report on the performance of FREME NER compared to other six NER systems on the same set of datasets. We report only the micro F1 score for weak and strong annotation match type of experiment.

The results show that for two datasets, MSNBC and KORE50, FREME NER achieved best performance. The results also show that for the RSS-500 and the Reuters-128 FREME NER achieved second best results, while for the DBpedia Spotlight dataset are achieved fourth best results. Note that we compared FREME NER to one of the most prominent NER systems such as DBpedia Spotlight, Babelfy, Entityclassifier.eu, FOX, NERD-ML and TagMe 2. Also note that we were not able to compute some scores for DBpedia Spotlight and TagMe 2 system, due to an unknown bug in those systems.

Table 2. Comparison of different NER systems and FREME NER.

Tool/Dataset	Exp. type	Spotlight	MSNBC	Reuters-128	KORE50	RSS-500
FREME NER	Weak	0.349	**0.914**	0.813	**0.956**	0.677
	Strong	0.242	**0.805**	0.675	**0.894**	0.579
DBpedia	Weak	0.413	0.559	0.512	n/a	0.422
spotlight	Strong	0.392	0.481	0.331	0.493	0.359
Babelfy	Weak	0.319	0.554	0.495	0.729	0.413
	Strong	0.250	0.470	0.310	0.690	0.277
Entityclassifier.eu	Weak	0.344	0.845	0.766	0.941	0.609
NER	Strong	0.256	0.683	0.553	0.879	0.535
FOX	Weak	0.222	0.348	**0.887**	0.833	**0.694**
	Strong	0.189	0.029	**0.618**	0.784	**0.618**
NERD-ML	Weak	**0.672**	0.632	0.484	0.760	0.391
	Strong	**0.564**	0.534	0.374	0.728	0.267
TagMe 2	Weak	0.663	0.454	n/a	0.766	0.521
	Strong	n/a	n/a	0.305	n/a	0.354

5   Chainable Web Services

As described previously, FREME NER is just one e-Service provided by the
FREME framework. A key benefit of FREME is that its pipelining approach
allows combination of e-Services. This capability will be explained with the
example in Listing 1.2.

```
 1  {
 2  "id": 55,
 3  "description": "Example pipeline",
 4  "serializedRequests": [
 5     {
 6     "endpoint": "http://api-dev.freme-project.eu/current/e-entity/freme-ner/documents",
 7     "parameters": {"language": "en"}
 8     },
 9     {
10     "endpoint": "http://api.freme-project.eu/current/e-link/documents/",
11     "parameters": {"templateid": "3"}
12     },
13     {
14     "endpoint": "http://api-dev.freme-project.eu/current/e-terminology/tilde",
15     "parameters": {
16     "source-lang": "en", "target-lang": "nl" }
17     },
18     {
19     "endpoint": "http://api-dev.freme-project.eu/current/e-translation/tilde",
20     "parameters": {
21     "source-lang": "en", "target-lang": "nl" }
22     }
23     ] }
```

Listing 1.2. Pipeline combining several e-Services.

A pipeline consists of one or more steps. All steps are embedded in the
serializedRequests JSON array. The order within the array defines the order of
execution. Each step has a mandatory service endpoint and, depending on the
endpoint, various optional or mandatory parameters. The steps can take various
input formats. If, like in the example, no format is specified, a step assumes NIF
input.

The first step in the example pipeline evokes FREME NER, which has been
described in the previous section. The second step uses the e-Link service to
retrieve information with a selected SPARQL query template. The template
used in the example[21] retrieves geospatial information. The third step calls the e-
Terminology service to enrich the content with terminology related information.
This step needs a source and a target language, here English and Dutch. The
last step calls the e-Translation service, with the same language pairs.

When services are chained, the chaining is controlled by the order of the
steps. In the example, FREME NER is followed by e-Link, which is followed by
e-Terminology and e-Translation. There is no separate workflow controller. The
data is sent between workflow steps without the need to explicitly interconnect
them. This approach greatly simplifies the work for authors of pipelines.

---

[21] See http://api.freme-project.eu/current/e-link/templates/3 to access the definition
of the template.

Each step can take the default processing format for FREME as input: text/turtle. It then processes the same format, without the need to explicitly declare the format for each step. For the first and the last step, that is, input and output of the pipeline, the pipeline author can declare formats explicitly. As of writing, HTML, XML, XLIFF 1.2 and ODT are accepted as input.

The formats are not declared in the pipeline service itself, but as an HTTP Content-Type header in the service request. FREME then calls the e-Internationalisation service to process the format. In that way, a pipeline can be re-used with all formats. As output for roundtripping, currently HTML is accepted.

The example pipeline shows several benefits. First, one can compare the outcome of several e-Services. In the example, named entity recognition and terminology annotation are used to enrich the same content. This combination has the potential to improve both services via data based comparisons.

Second, there is no need to hardwire the combination of services, as long as the services adhere to the linguistic linked data stack. This can be seen in line 10 of the example. The e-Link service is installed on a different server (with the domain api.freme-project.eu) than the other e-services. The combination of services does not need a hardwired integration.

Third, the pipeline and in this way the e-Services are agnostic to given input and output formats. Format coverage is realised via the previously described e-Internationalisation service. Separating the actual services and the formats to be processed has the advantage that other services can easily be integrated and benefit from the growing set of formats being supported. The example in Listing 1.3 shows how to call a pipeline with two alternative HTTP requests, executed via cURL.

```
1  curl -X POST -H "Content-Type: text/plain" -d 'Berlin is a nice city.'
2  "http://api.freme-project.eu/current/pipelining/chain/1"
3  curl -X POST -H "Content-Type: text/xml" -d 'Berlin is a nice city.'
4  "http://api.freme-project.eu/current/pipelining/chain/1"
```

Listing 1.3. Example CURL request.

The only difference between the requests is the Content-Type header. In the second request, it is set to XML, which allows processing of general XML content.

If both input and output are set to the content type text/html, roundtripping becomes possible. That is, the enrichment information is stored in the actual HTML content. An example is given in Listing 1.4. Here, a pipeline of first e-Entity, then e-Terminology has been applied to the HTML content. The resulting HTML contains dedicated attributes to store the term and entity related information.

```
1  <p>Welcome to the <span its-term-info-ref="http://example.com/#char=36,40"
2  its-term="yes">city</span> of <span
3  its-ta-class-ref="http://dbpedia.org/ontology/Settlement"
4  its-ta-ident-ref="http://dbpedia.org/resource/Prague">
5  Prague</span>.</p>
```

Listing 1.4. Pipeline with roundtripping of HTML.

The combination of roundtripping and several e-Services has the potential to contribute to a data-driven comparison of e-Service outputs. In the example, there is information available from structured (HTML) markup, e-Terminology and e-Entity in the output. This representation makes a query trivial like: find all instances of entities which are in the same markup context as certain terms. The paragraph, represented as markup via the p element, would be a result for such a query, with the term city and the entity Prague. Such an interrelation of NLP output and original markup is not the aim of the current paper, but an interesting future topic for research.

Fourth, the pipelining greatly allows for automatization of repetitive processes and for making the content itself intelligent. For example, a client application could analyze the content with regards to the language of content and use this information for adapting the pipeline automatically.

6 Conclusion

This paper introduced the current state of the FREME framework with regards to two aspects: named entity recognition via the FREME NER e-Service, and chaining of e-Services. In addition, we have put FREME into the context of linguistic linked data and related approaches of multilingual and semantic processing.

The discussion on FREME NER showed some preliminary evaluation results. The pipelining of e-Services has a practical benefit (e.g. ease and automatization of similar language and data processing workflows), but also a research potential. The combination of named entity recognition, terminology annotation and machine translation can lead to a data driven improvement of all of these technologies. This is a potential next step for FREME.

Acknowledgement. This work was supported via the FREME project funded from the EU Horizon 2020 research and innovation programme under grant agreement No 644771.

References

1. Auer, S., Bizer, C., Kobilarov, G., Lehmann, J., Cyganiak, R., Ives, Z.: DBpedia: a nucleus for a web of open data. In: Aberer, K., et al. (eds.) ASWC/ISWC - 2007. LNCS, vol. 4825, pp. 722–735. Springer, Heidelberg (2007). doi:10.1007/978-3-540-76298-0_52
2. Bachmann-Gmur, R.: Instant Apache Stanbol. Packt Publishing Ltd. (2013)
3. Brümmer, M., Dojchinovski, M., Hellmann, S.: DBpedia abstracts: a large-scale, open, multilingual NLP training corpus. In: Calzolari, N., et al. (eds.) Proceedings of the Tenth International Conference on Language Resources and Evaluation (LREC 2016). European Language Resources Association (ELRA), Paris, France, May 2016
4. Fielding, R.T., Taylor, R.N.: Principled design of the modern web architecture. ACM Trans. Internet Technol. (TOIT) **2**(2), 115–150 (2002)

5. Finkel, J.R., Grenager, T., Manning, C.: Incorporating non-local information into information extraction systems by Gibbs sampling. In: Proceedings of the 43rd Annual Meeting on Association for Computational Linguistics, pp. 363–370. Association for Computational Linguistics (2005)
6. Hellmann, S., Lehmann, J., Auer, S., Brümmer, M.: Integrating NLP using linked data. In: Alani, H., et al. (eds.) ISWC 2013. LNCS, vol. 8219, pp. 98–113. Springer, Heidelberg (2013). doi:10.1007/978-3-642-41338-4_7
7. Mendes, P.N., Jakob, M., García-Silva, A., Bizer, C.: DBpedia spotlight: shedding light on the web of documents. In: Proceedings of the 7th International Conference on Semantic Systems, pp. 1–8. ACM (2011)
8. Pautasso, C., Zimmermann, O., Leymann, F.: Restful web services vs. Big'Web services: making the right architectural decision. In: Proceedings of the 17th International Conference on World Wide Web, pp. 805–814. ACM (2008)
9. Sasaki, F., et al.: Introducing FREME: deploying linguistic linked data. In: Proceedings of the 4th Workshop on the Multilingual Semantic Web (2015)
10. Usbeck, R., Röder, M., Ngomo Ngonga, A.-C.: Evaluating entity annotators using GERBIL. In: Gandon, F., Guéret, C., Villata, S., Breslin, J., Faron-Zucker, C., Zimmermann, A. (eds.) ESWC 2015. LNCS, vol. 9341, pp. 159–164. Springer, Cham (2015). doi:10.1007/978-3-319-25639-9_31
11. Usbeck, R., et al.: GERBIL: general entity annotator benchmarking framework. In: Gangemi, A., Leonardi, S., Panconesi, A. (eds.) Proceedings of the 24th International Conference on World Wide Web, WWW 2015, Florence, Italy, 18–22 May, pp. 1133–1143. ACM (2015)

Entity Typing Using Distributional Semantics and DBpedia

Marieke van Erp$^{(\boxtimes)}$ and Piek Vossen

Vrije Universiteit Amsterdam, Amsterdam, The Netherlands
{marieke.van.erp,piek.vossen}@vu.nl

Abstract. Recognising entities in a text and linking them to an external resource is a vital step in creating a structured resource (e.g. a knowledge base) from text. This allows semantic querying over a dataset, for example selecting all politicians or football players. However, traditional named entity recognition systems only distinguish a limited number of entity types (such as Person, Organisation and Location) and entity linking has the limitation that often not all entities found in a text can be linked to a knowledge base. This creates a gap in coverage between what is in the text and what can be annotated with fine grained types.

This paper presents an approach to detect entity types using DBpedia type information and distributional semantics. The distributional semantics paradigm assumes that similar words occur in similar contexts. We exploit this by comparing entities with an unknown type to entities for which the type is known and assign the type of the most similar set of entities to the entity with the unknown type. We demonstrate our approach on seven different named entity linking datasets.

To the best of our knowledge, our approach is the first to combine word embeddings with external type information for this task. Our results show that this task is challenging but not impossible and performance improves when narrowing the search space by adding more context to the entities in the form of topic information.

1 Introduction

Fine grained entity typing facilitates precise queries to structured datasets. It can, for example, be used to query for all politicians or presidents in a dataset. With natural language processing techniques (NLP) becoming more accurate, structured datasets are increasingly being generated from text. However, there is still a gap between the results generated by most NLP techniques and what semantic web resources can offer.

Named entity recognition and classification (NERC) systems usually only discern a limited number of coarse grained types such as person, location, organisation and miscellaneous (CoNLL, [27]) or person, organisation, location, facility, weapon, vehicle and geo-political entity (ACE, [1]).

To obtain fine grained entity types for an entity, a named entity linking step is often employed to link recognised entities in an existing knowledge base such

© Springer International Publishing AG 2017
M. van Erp et al. (Eds.): ISWC 2016 Workshops, LNCS 10579, pp. 102–118, 2017.
https://doi.org/10.1007/978-3-319-68723-0_9

as DBpedia. Thereby linked entities are enriched with the types of the resource, resolving the problem of not being able to perform fine grained queries. However, entity linking does not solve the entire problem, as not all entities can be linked to the knowledge base, for example, because there is no suitable resource present (often denoted as 'NIL' entities) or the resource may not contain any useful information about the entity to facilitate semantic querying [5].

In this paper, we focus on predicting the entity type of an entity regardless of its presence or absence in the knowledge base. Once the entity type has been established, a schema can be assigned to an entity which can serve as input for identifying other characteristics of the entity for example in a knowledge base population task. To tackle this task, we present an approach that employs distributional semantics and DBpedia types. We evaluate our approach on seven different entity linking benchmark datasets and make our resulting datasets available as a first dataset of NIL entities with fine grained type information. The contributions of this paper are threefold:

1. a method and implementation for fine grained entity typing;
2. quantitative and qualitative evaluation and analysis of the system on seven benchmark datasets; and
3. a new dataset for NIL entities including entity types.

The remainder of this paper is organised as follows. In Sect. 2, background and related work is discussed. Our approach is presented in Sect. 3. Section 4 describes the resources and datasets used for the experiments presented subsequently in Sect. 5. An analysis of the results (Sect. 6) and conclusions and future work (Sect. 7) wrap up this paper. All code, links to datasets and experiments are available via https://github.com/MvanErp/entity-typing.

2   Background and Related Work

Named entity recognition and classification has a long tradition in the natural language processing field, starting with the Message Understanding Conferences that were organised by DARPA between 1987 and 1997 [7]. The field also received a boost with the CoNLL 2002 and 2003 named entity recognition shared tasks [24, 27], whose annotated datasets are still widely used for training and testing named entity recognition and classification approaches. However, the entity types used in these shared tasks and in the ACE challenges [1] are quite limited; CoNLL only distinguishes four entity types, and ACE seven main types as well as some subtypes. The main reason for this is that most systems developed for these tasks rely on supervised machine learning, for which sufficient examples of each entity type are needed. Some experiments with more elaborate type hierarchies [25] and (semi-)supervised machine learning for fine grained entity typing have been carried out [18], but this has not caught on much. Most likely due to the prohibitive expense of creating training datasets for this.

Named entity disambiguation or named entity linking systems can implicitly provide fine grained entity types. With Wikipedia and later DBpedia generic and large resources were for the first time widely available. [13] present a system that uses the Wikilinks to link Wikipedia articles to relevant keywords in a text. [16] present the first entity linking system. Their system identifies entities in a text by a machine learning algorithm and then links them to a Wikipedia article, augmenting the text with the Wikipedia article's category information. For an overview of more entity linking approaches using Wikipedia see [8].

In the Semantic Web community entity linking systems such as [12,28] rely on the knowledge base providing a good coverage of the entities in a text to link. If the entity in the text does not have a suitable resource in the knowledge base, the system returns a NIL value at best, and at worst an incorrect link.

Approaches that deal with NILs or can be considered entity typing without entity linking are found in the domain of entity clustering [4]. However, these approaches generally do not leverage type information or hierarchies from external resources. Clustering methods and distributional models such as word2vec (which will be further explained in Sect. 4) have in common that they utilise the context surrounding a word or entity. Thus far, distributional models have been used for a wide variety of natural language processing tasks including relationship learning [19] and named entity recognition [26], but to the best of our knowledge, our work is the first to apply it to the entity typing.

In the remainder of the paper, we make a distinction between entity mentions, i.e. the textual reference in a text that denote entities where entities are things in the real world, oftentimes denoted by URIs in a knowledge base, e.g. a DBpedia resource. Entity types are properties of entities that express a categorisation of the entity. An example of two entity mentions referring to the same entity would be "Royal Air Force" and "RAF" both referring to http://dbpedia.org/resource/Royal_Air_Force.

3 Entity Typing Using Distributional Semantics

Our approach for entity typing relies on the assumption that similar words[1] occur in similar contexts. As entities may not be mentioned often in a text, a large corpus of texts is also needed.

For the word embeddings in the model, we use word2vec [14][2], a popular word embeddings implementation. The idea behind word vectors is not new [2], but Mikolov et al. propose two new optimised architectures, in addition to a freely available implementation of the algorithm, making it possible to perform experiments with decently-sized datasets on fairly standard machines.[3]

The approach relies on a neural network to learn word vectors from a large text corpus, the idea being that similar words occur in similar contexts which can be captured by a word vector model. Such a model can be used to compute

---

[1] As entities are made up of words, we hypothesise that this also extends to entities.

[2] https://code.google.com/archive/p/word2vec/.

[3] For this paper, we ran experiments on a Ubuntu machine with 2 CPUs, 16 GB of RAM and most experiments did not take longer than 2 h.

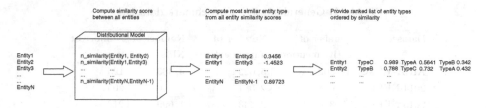

Fig. 1. Entity typing using distributional semantics system setup

the semantic distance between two words or phrases, as well as algebraic operations on the vectors, an often mentioned example here being *vector("King") - vector("Man") + vector("Woman")* resulting in *vector("Queen")* being the closest response [15]. It also allows to compute a measure of semantic similarity between two words or groups of words which is what we will employ here. One expects the similarity between the entities 'George Bush' and 'Barack Obama' to be higher than between the entities 'George Bush' and 'Mexico City', which is indeed what the GoogleNews model returns (0.50 vs. 0.15). By performing this computation over many entity pairs and aggregating the entity types of the most similar entities, we aim to assign a meaningful entity type to an entity for which the entity type is unknown.

Figure 1 shows the system setup. A list of entity mentions for which the entity type is not known serves as input to the system. This entity mention is compared to all other entity mentions in the dataset for which the type information is available and the similarity between the two entities is computed. Thus, we average the scores of all entity types and produce a ranked list of the entity types whose entities are most similar to the entity at hand.

By using entity linking benchmark datasets that contain links to DBpedia, the entity types of the entities can be retrieved. Furthermore, the seven different datasets that will be described in the next section provide a wide range of entity mentions and types to evaluate various aspects of the approach.

Two types of experiments will be carried out: **1. Dataset-based experiments:** in this series of experiments, every entity mention within a dataset will be compared to all other entity mentions in that dataset. **2. Topic-based experiments:** in this series of experiments, the dataset is first split into topics and every entity mention is compared to only those entity mentions that are also in the same topic.

4   Resources and Datasets

The experiments carried out for this paper rely on existing algorithms and datasets which are described here.

4.1   Benchmark Entity Linking Datasets

We chose to test our approach on a number of freely available entity linking benchmark datasets, and were previously collected and described in [6]. Each of

Table 1. General statistics benchmark datasets

Dataset	Number of entity mentions	Number of unique entities	Number of NILs	Number of types
AIDA-YAGO	11,862	5,029	4,333	195
2014 NEEL	3,084	2,081	0	205
2015 NEEL	5,346	2,643	1,699	213
OKE2015	773	501	116	55
RSS500	874	369	449	97
WES2015	9,753	6,016	1	211
Wikinews	1,724	251	660	48

these datasets has different characteristics, that may present our approach with different challenges, all have in common that are linked to DBpedia enabling us to leverage the type information from DBpedia.[4]

Table 1 displays some general statistics on the datasets. For the RSS500 dataset and WES2015, entities that did not have a DBpedia link were counted as NILs (these were linked to a dataset specific resource, but the type information for these was not available). In 2014 NEEL, NILs are not annotated.

AIDA-YAGO2 Dataset

The AIDA-YAGO2 dataset [9][5] is an extension of the most commonly used named entity recognition benchmark dataset, namely the CoNLL 2003 entity recognition task dataset [27]. The CoNLL 2003 dataset is based on a 10-day subset of Reuters news articles published between August 1996 and August 1997 by Reuters, to which part-of-speech and chunk tags were added automatically and named entities were added manually.

For this paper, we have mapped the Wikipedia URL to its corresponding DBpedia URI. Furthermore, the Reuters topic descriptions were reinserted into the articles in order to perform a series of experiments with topic classifications. The majority of the codes was added semi-automatically by Reuters; first a rule-based system proposes a topic, this is then checked by one or two human annotators. Next, the topic codes were cleaned up and its ancestors in the hierarchy were added through the process described in [11], whose corrected dataset we used.[6]

---

[4] AIDA-YAGO2 originally contained Wikipedia URLs but these have been mapped to their corresponding DBpedia URIs.

[5] https://www.mpi-inf.mpg.de/departments/databases-and-information-systems/research/yago-naga/aida/downloads/.

[6] Available from: http://www.jmlr.org/papers/volume5/lewis04a/lyrl2004_rcv1v2_README.htm Last visited: 27 April 2016.

2014 and 2015 NEEL

The 2014 and 2015 Named Entity rEcognition and Linking (NEEL) dataset is made up of two Twitter datasets used in two consecutive challenges. The 2014 NEEL dataset [3][7] consists of 3,504 tweets extracted from over 18 million tweets provided by the Redites project. The tweets were collected over a period of 31 days between 15 July 2011 and 15 August 2011 and include noteworthy events. The 2014 Microposts challenge dataset was created to benchmark automatic extraction and linking entities.

The 2015 NEEL corpus [21][8] is an extension of the 2014 dataset containing 6,025 tweets. Additional tweets published in 2013 were added to the original dataset. The resulting corpus was further extended to include entity types and NIL references. Entity references are linked to DBpedia resources.

OKE2015

The Open Knowledge Extraction Challenge 2015 (OKE2015) [20][9] corpus consists of 197 sentences from Wikipedia articles. Besides entities linked to DBpedia, the entities are also annotated with Dolce Ultra Lite classes[10], coreference relations, and semi-automatic anaphora resolution, and detection of emerging entities. The corpus was split into a train and test set containing 96 sentences for the training set, and 101 for the test set.

RSS-500-NIF-NER

The RSS-500 dataset [23][11] contains data from 1,457 RSS feeds, including major international newspapers, covering a wide variety of topics. 500 sentences were chosen from an initial corpus of 11.7 million sentences and annotated by one researcher. The chosen sentences contain a formal relation (e.g. "..who was born in.." for dbo:birthPlace), that should occur more than 5 times in the 1% corpus.

WES2015

The WES2015 dataset was originally created to benchmark information retrieval systems [29].[12] The documents originate from a blog about history of science, technology, and art.[13] in which the entities are linked to DBpedia resources. The dataset also includes 35 annotated queries inspired by the blog's query logs, and relevance assessments between queries and documents. These were not used in the experiments described in this paper.

---

[7] http://scc-research.lancaster.ac.uk/workshops/microposts2014/challenge/index.html.

[8] http://scc-research.lancaster.ac.uk/workshops/microposts2015/challenge/index.html.

[9] https://github.com/anuzzolese/oke-challenge.

[10] http://stlab.istc.cnr.it/stlab/WikipediaOntology/.

[11] https://github.com/AKSW/n3-collection.

[12] http://yovisto.com/labs/wes2015/wes2015-dataset-nif.rdf.

[13] http://blog.yovisto.com/.

WikiNews/MEANTIME

The WikiNews/MEANTIME (hereafter referred to as 'Wikinews') [17].[14] is a linguistically and semantically annotated corpus of 120 news articles from the open news website Wikinews.[15] This corpus is divided into four sub-corpora: Airbus, Apple, General Motors and Stock Market. These are annotated with entities in text, including links to DBpedia, events, temporal expressions and semantic roles. This set of articles was selected to represent domain entities and events from the financial aspect of the automotive industry. The corpus is available in English, Spanish, Italian and Dutch. In our experiments, we limit ourselves to the English part of the corpus.

4.2 Word2vec Models

In this paper, we use three different word2vec models, the first two are pre-trained models: (1) GoogleNews-vector-negative300.bin.gz[16] trained on part of the Google News dataset ($\sim$100 billion words) [15][17] and (2) English Wikipedia (Feb 2015).[18] The third model was generated from the Reuters RCV1 corpus,[19] consisting of news wire published between August 1996 and August 1997. One of the main entity linking datasets described in the next section is derived from this dataset, therefore we chose to do an experiment involving this dataset and compare it to the Google News corpus. For all experiments, we use the Python gensim implementation of word2vec.[20]

5   Experiments and Results

The system outputs a ranked list of entity types for each entity mention. When measuring the performance of the system against the gold standard entity type, the precision at positions 1, 5 and 10 in the ranked results list is measured.

The results are divided into coarse- and fine-grained results. For the coarse-grained results, we only looked at the top level entity types in DBpedia, e.g. Agent, Place, Name, TopicalConcept etc. For the fine-grained results, we only looked at the most specific types in the DBpedia ontology. e.g. Airline, BeautyQueen, Monastery etc. This is quite a strict evaluation metric as we only either the most generic or most specific exact entity type per entity. If a system returns entity types from elsewhere in the type hierarchy, these are not currently not

---

[14] http://www.newsreader-project.eu/results/data/wikinews.
[15] https://en.wikinews.org/.
[16] https://drive.google.com/file/d/0B7XkCwpI5KDYNlNUTTlSS21pQmM/edit? usp=sharing.
[17] Unfortunately, no further information about the Google News corpus is available as it is not an open dataset.
[18] https://github.com/idio/wiki2vec.
[19] http://trec.nist.gov/data/reuters/reuters.html.
[20] https://radimrehurek.com/gensim/models/word2vec.html.

Table 2. Results per entity dataset as percentage of correct types returned by the system in position 1, 5 or 10.

Dataset	GoogleNews						Wikipedia						Reuters					
	Coarse			Fine			Coarse			Fine			Coarse			Fine		
	@1	@5	@10	@1	@5	@10	@1	@5	@10	@1	@5	@10	@1	@5	@10	@1	@5	@10
AIDA-YAGO2	0.00	0.23	0.42	0.26	5.71	16.62	0.00	0.19	0.38	0.38	4.33	14.06	0.00	0.12	0.25	0.40	5.86	16.24
2014 NEEL	0.00	1.53	3.45	0.13	5.62	12.48	0.00	1.25	3.26	0.04	4.99	12.42	0.00	0.76	2.93	0.05	2.42	9.18
2015 NEEL	0.00	1.77	3.05	0.04	5.60	12.79	0.00	1.57	2.69	0.08	4.64	12.40	0.00	0.53	1.32	0.05	1.95	6.37
OKE2015	0.00	0.84	1.05	1.68	7.58	16.84	0.00	0.76	1.33	2.66	7.02	17.27	0.00	0.00	1.55	4.88	9.98	20.40
RSS500	0.00	0.12	0.25	6.37	9.36	15.23	0.00	0.00	0.27	4.28	7.35	12.83	0.00	0.00	0.33	0.17	2.97	7.26
WES2015	0.00	0.61	1.91	0.10	2.19	5.41	0.00	0.97	3.60	0.01	5.51	9.48	0.00	1.02	3.25	0.05	3.57	6.74
Wikinews	0.00	8.39	16.46	1.90	11.87	28.48	0.00	16.52	22.71	2.32	8.77	26.71	0.00	6.00	12.69	1.91	8.05	25.24

Table 3. Statistics on coverage of entities in the different models

	Total #	GoogleNews		Wikipedia		Reuters	
	Entity mentions	Found	Not found	Found	Not found	Found	Not found
AIDA-YAGO2	11, 862	9, 103	2,759 (23.26%)	8, 294	3,568 (30.07%)	8, 937	2,925 (24.66%)
2014 NEEL	3,084	2,347	737 (24.18%)	2,487	597 (19.36%)	1,982	1,102 (35.73%)
2015 NEEL	5,346	2,949	2,397 (44.83%)	2,885	2,461(46.03%)	2,101	3,245 (60.70%)
OKE2015	773	554	219 (28.33%)	620	153(19.79%)	529	244 (31.57%)
RSS500	874	801	73 (8.35%)	748	126(14.41%)	606	268 (30.66%)
WES2015	9, 753	6, 743	3,010 (30.86%)	8, 278	1,475 (15.12%)	6, 496	3,257 (33.40%)
Wikinews	1, 724	985	739 (42.86%)	1, 311	413 (23.96%)	1, 265	459 (26.62%)

considered in the aggregated results, but these will be discussed in the qualitative analysis in Subsect. 6.3.

Tables 2 and 3 present the results of the experiments of the dataset-based experiments. The first table displays the percentage of correct entity types returned by the system in the ranked list at positions 1, 5 and 10. Table 3 provides statistics on the coverage of the entity mentions in the various word2vec models.

For the AIDA-YAGO and Wikinews datasets, a topic classification of the articles from which the entities are derived is available too. In this subsection, a series of experiments is described in which the entity datasets are further divided into datasets by topic. As this results in fewer entity comparisons (as only entities within a topic are compared), this narrows down the search space.

As described in Sect. 4, we inserted the Reuters topics classification into the AIDA-YAGO dataset. Two series of experiments were run: one with only the top level Reuters topics (AIDA-YAGO Coarse) and one with the more fine grained Reuters topics (AIDA-YAGO Fine). In total, the Reuters topics classification set contains top level 23 topics, but only 21 are present in the dataset. In total, the Reuters hierarchy contains 103 subtopics, of which 68 are present in the dataset. One article does not have a topic ascribed to it, this article was treated as a separate topic resulting in 22 topics in total in the coarse grained top-level Reuters topics and 69 topics in the finer grained topics experiments. An overview of the topics and their distribution over the AIDA-YAGO dataset can be found on our github page.

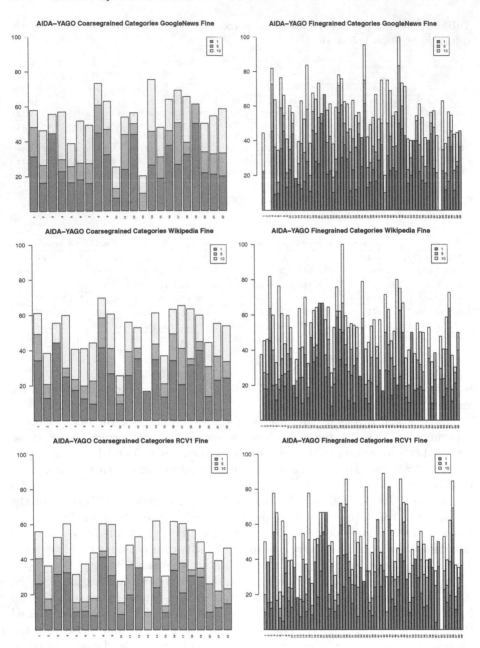

Fig. 2. Results on AIDA-YAGO dataset with entities divided into topics

The entities in the Wikinews topics are fairly evenly spread, the largest topic contains 275 unique entities and the smallest 104, with a median at 140. The AIDA-YAGO topics are quite diverse in nature and in division; in the fine grained

Table 4. Results per entity dataset with aggregated topics

Dataset	GoogleNews						Wikipedia						Reuters					
	Coarse			Fine			Coarse			Fine			Coarse			Fine		
	@1	@5	@10	@1	@5	@10	@1	@5	@10	@1	@5	@10	@1	@5	@10	@1	@5	@10
AIDA-YAGO Coarse	0.00	0.27	6.42	0.02	7.80	22.28	0.00	0.30	5.29	0.04	6.73	20.24	0.00	0.16	2.83	0.02	7.04	21.60
AIDA-YAGO Fine	0.00	0.31	10.37	0.30	11.64	29.61	0.00	0.45	10.08	0.18	10.01	27.06	0.00	0.24	6.92	0.27	9.57	25.40
Wikinews Topics	0.00	4.23	14.37	0.28	3.10	9.01	0.00	3.35	11.51	0.21	2.09	6.49	0.00	0.00	6.36	3.64	7.50	10.23

topic division, the smallest topics only contains 7 unique entity mentions (Personal Income) and (Reserves), and the largest 18,616 (Sports). The median lies around 108 entity mentions per topic. Even on the more coarse grained topic division the differences are large: the smallest topic contains 29 entity mentions (Management), whereas the largest contains 30,320 (Government/Social), but the median lies around 295 entity mentions per topic.

The Wikinews dataset is divided into four themes, by treating the entities in each of these themes, we can also experiment with a really high-level topic classification. Figure 2 and Table 4 show the results of the experiment series in which entity linking was performed within a topic. For space considerations, only the results on the AIDA-YAGO dataset are presented here, the full results can be found on the https://github.com/MvanErp/entity-typing.

In Fig. 2, the graphs on the lefthand side show the results on the coarse grained Reuters topics, and the righthand side show the results on the fine grained Reuters topics. The figures shown here only present the results on the most specific DBpedia types (columns 'Fine' in Table 4).

6   Results Analysis and Discussion

In this section, we first present a quantitative analysis of the dataset- and topic-based experiments, followed by a qualitative analysis of a data sample.

6.1   Dataset-Based Experiments

At first sight, the results may indicate that the approach does not yield the desired results for entity typing but there are two main reasons that indicate otherwise. Firstly, the smaller datasets OKE2015 and Wikinews perform better than the larger datasets. When considering the entities in these datasets, they do not only cover fewer different entities, but also fewer different entity types (see Table 1). The topic-based experiments demonstrate that datasets that are less broad, i.e. centred around a particular topic are better suited to our approach. Secondly, to preserve coherence, we only focused on type information from the

DBpedia ontology (http://dbpedia.org/ontology/) but not all resources contain a DBpedia ontology type, which will be discussed in Subsect. 6.3.

Let us first consider the differences in results between the different word2vec models. The GoogleNews model is almost 10x larger than the Reuters model (3.4 GB vs 374 MB) and for the majority of the datasets tested, it yields a better performance, due to there being more contexts in the model for a particular entity. However, as Table 2 shows, this does not hold for the AIDA-YAGO dataset, whose entities are extracted from the Reuters corpus. The results show that there is a clear advantage to using a model based on the data the entities are derived from, as this ensures that the original context in which the entities were mentioned are also encoded in the model.

Another interesting observation is that the approach consistently ranks more fine grained entity types first over more coarse grained types. This accounts for the coarse grained Score@1 columns yielding scores of 0.00, whilst the approach manages to find the most fine grained entity type (according to the gold standard) in the top position in some cases. This is a positive signal as the goal is to provide fine grained entity types. It should be noted here that intermediate entity types are not taken into account in this analysis, e.g. in the type hierarchy Place - PopulatedPlace - Settlement - Village, the coarse grained evaluation measured whether the type 'Place' was returned by the system, and the fine grained analysis whether the type 'Village' was returned by the system. As the hierarchy does not have a fixed number of levels, measuring the performance at each step is difficult to aggregate and is left for future work.

As Table 3 indicates, the corpus used to find and compare entities is an important factor in the experiments. Not surprisingly, the entities in RSS500 are largely covered by the GoogleNews model, with only 8.35% of the entity mentions missing, but this figure jumps to 30.66% for the Reuters model. Both corpora cover news, but the Reuters corpus dates from nearly 20 years ago, when different entities played a role in the news. This is also apparent from the coverage on the 2014 and 2015 NEEL datasets in Reuters, and to a lesser extent in GoogleNews and Wikipedia. Tweets generally have a different style than news with more abbreviations, capitals and hash tags and Twitter handles. As the models and entity mentions are not normalised, this yields fewer matches.

As the OKE2015 dataset is based on Wikipedia, the coverage by the Wikipedia model on this dataset is higher than that of the news models. The coverage of the WES2015 dataset is also better in the Wikipedia model than the news models, this is due to the WES2015 dataset covering science topics. Mentions of pre-socratic philosophers such as 'Anaxagoras' and 17th century botanists such as 'Nicholas Culpeper' are simply less frequent in the news than in sources such as Wikipedia. In the word2vec models, there is also a slight bias towards more frequent entities, it would simply not be possible to capture all hapaxes as this makes the model less efficient. In training the Reuters model, only words that occur at least 10 times in the corpus were taken into account, as the same parameters were used as those for the GoogleNews model. Further experiments with different parameter settings may also positively influence the performance.

6.2   Topics-Based Experiments

When looking at the fine grained results, on the righthand side of Fig. 2, a few topics yield scores of 0. Topics 1 (Advertising Promotion), 58 (Reserves), and 62 (Share listings) prove difficult for the Wikipedia model to classify correctly. These topics contain 18, 7 and 14 entity mentions respectively. However, Topic 40 (Leading indicators), with only 20 entity mentions, does yield reasonable scores. Upon inspecting the entities in this topic, this is probably due to these being fairly generic entities such as 'Hungary', 'Budapest' and 'Spain', these entities occur so often that only a few contexts suffice to type them. Interestingly, the GoogleNews model also has trouble with Topic 3 (Art Culture Entertainment), which contains 256 entity mentions. As the entities are derived from the Reuters corpus, it performs decently there, but also the Wikipedia model does well here as it contains a fair amount of information regarding entertainment [10]. Topics 21 (EC External Relations) and 22 (EC Monetary Economic) with 57 and 12 entities perform well across all models. Topics 29 (Fashion) and 48 (Money Supply) obtain a 100% score on the top 10 results in the Wikipedia and GoogleNews models respectively, providing a hunch about the coverage of the model.

For the coarse grained topics, on the lefthand side of Fig. 2, the size of the topic plays a role again. Topic 13 (Management) is the smallest topic, with only 29 entity mentions, and it yields the lowest results in all models. However, bigger is not always better, as Topic 10 (Government/Social) with 30,320 entity mentions confirms.

This topic contains sports as well as subtopics such as obituaries, elections, weather, crime and fashion. The topics that stand out positively here are 3 (Consumer finance) and 19 (Output capacity) with 37 and 29 entity mentions, respectively, and 4 (Contracts/Orders), 8 (European Community) and 14 (Markets/Marketing) with 105, 279 and 407 entity mentions respectively.

The figures for the more coarse grained DBpedia types can be found on the Github page. These are largely in line with the fine grained results where the same topics present the different models with challenges. As with the overall dataset results, the system aims for more specific entity types first, and less specific types are typically found in the rank 5–10 range and hardly any coarse grained entity types are found at rank 1.

The results for the Wikinews dataset do not improve much in the per-topic setting. This dataset is not that large, and still quite broad. Another issue in this dataset is that there are many entities that are not in DBpedia, thus resulting in fewer entities with context typing, this is partly due to the wide variety in entities accepted in the annotations such as 'GM's package' and 'JSF contract'. The fine grained DBpedia type results on the Wikinews dataset are a bit 'all or nothing' in the sense that the majority of the correct results are found at rank 1, proportionally fewer correct results are found lower in the ranked list. As with the overall dataset results, the system aims for more specific entity types first, and less specific types are typically found in the rank 5–10 range.

6.3   Qualitative Analysis

For the qualitative analysis we looked at various samples of the data to get a more in-depth understanding of the mismatch between the system output and the gold standard types.

Not Matching the Gold Standard. We took a random sample of 200 entries dataset-based experiments for which no match was found between the gold standard types and the entity types proposed by the system. There are four interesting observations here:

1. Multifaceted entities: Some entities are quite difficult to capture in a single DBpedia type, e.g. http://dbpedia.org/resource/Thomas_Erle is classified as a MilitaryPerson in the gold standard. He was indeed an army general, but later a politician who sat in the House of Commons and our system returns type suggestions such as Politician, MemberOfParliament and Congressman. Other ambiguous entities are found when the entity mention can have several meanings. E.g. "Buffalo" is linked to http://dbpedia.org/resource/Buffalo,_New_York which is of course a City, but the system returns types such as Eukaryote, Species and Animal. To resolve these, more context may be included in the query.

2. Type Specificity: As the system favours more specific entity types over more generic ones, it may in some cases suggest a type that is also correct, but was not present in the gold standard. For example http://dbpedia.org/resource/Justin_Bieber has type Person in DBpedia, and the system suggests MusicalArtist.

3. Ambiguous entity mentions: Some entity mentions are very difficult to classify without additional context. The entity "massacre" is linked to http://dbpedia.org/resource/The_Massacre which is a music album in DBpedia. However, the system predicts types such as MilitaryConflict and Event. Other difficult entities to type for the system are thing such as 'month' and numbers, which are annotated in some of the datasets. For such 'generic' terms denoting entities, the distributional semantics approach is too coarse and in some cases it is debatable whether the item should have been annotated as an entity at all.

4. Gold standard limitations: No dataset is perfect and due to their size it is impossible even for the very active Wikipedia and DBpedia communities to check every resource but there are some types in the gold standard that are at least a bit puzzling such as http://dbpedia.org/resource/KFC which is a SportsTeam according to the gold standard and the http://dbpedia.org/resource/US_Open_ (tennis) which is a PopulatedPlace.

No Gold Standard DBpedia Types Available. Some entity mentions have a DBpedia link assigned to them, indicating that they are present in the knowledge base, but the gold standard does not contain any DBpedia types to evaluate the system output against. There are two main reasons for this.

1. Choice of ontology: For reasons of coherence and manageability of the evaluations, only DBpedia ontology types were considered in our experiments. But DBpedia resources may also have Yago, Umbel, Geonames, Schema, and Wikidata types assigned to them. Although the majority of the resources in our datasets have at least one DBpedia type assigned, there are also resources that for example only have Yago types (e.g. http://dbpedia.org/resource/BRIC).

2. Redirects: The entity mention 'China' is linked to http://dbpedia.org/resource/People's_Republic_of_China which only contains yago types. However, this resource actually also has a redirect link to http://dbpedia.org/page/China which does contain the DBpedia types `Place` and `Country` for which were also suggested by the system.

Redirects and other ontologies were not considered in these experiments, but follow-up research could include these.

NILs. To gain insights in the performance of the approach on NIL entities, we manually evaluated the performance of the Reuters model on 200 random entity mentions from the AIDA-YAGO dataset with fine grained Reuters classes. As this is not a formal annotation, we only considered whether a 'reasonable' entity type was suggested by the system at rank 1, or within ranks 5 or 10, but we did not specify whether this was a fine grained or coarse grained type.

In 31 of the cases, our method suggests a reasonable entity type at position 1 in the ranked list. In an additional 60 cases, a relevant entity type was found within the first 5 results, and in another 20 cases within the first 10 positions. As there seem to be many Cricketers in the dataset, these tend to be classified correctly. The system also returns some very specific correct entity types in certain cases such as `BusCompany` for Swebus AB.

As the system favours more specific entities, it often comes up with various different types of sports teams (`RugbyTeam`, `SoccerTeam`) or athlete types (`TennisPlayer`, `RugbyPlayer`), so the suggested entity types are in the correct domain. Some entity mentions are very difficult to type, such as "Ontario-based", "US-led" and "non-EU". There are also some mentions of divisions of companies such as "Sydney Newsroom". Whilst strictly speaking not a company in itself, it is part of a company and thus deemed reasonably typed by the system.

There are also some cases where the entity mention boundary was probably not correct in the dataset such as "British Airways-American", although the system does return `Airline` as a type suggestion.

The analysis shows that the system suggests very reasonable entity types for the NILs in these datasets. An inspection of a number of topics and suggested types shows that the topic boundaries effectively limit the number of entity types for a topic, which is very helpful to the system as it provides a stronger context to type entities from. The typed NILs datasets generated in these experiments are available through our website.

The analyses described in this section indicate that some of the system suggestions are more reasonable than the results in Table 2 suggest.

7   Conclusion and Future Work

In this paper, we presented a novel method and experiments for fine grained entity typing using distributional semantics and DBpedia. Our results show that this is a difficult task but when entity mentions are limited within a topic, the system achieves reasonable performance. We tested this topic-based approach on two datasets with available topic information. In future experiments, we aim to use topic detection to investigate different topic granularity levels on all datasets to gain insights into the optimal topic-entity type-entity mention ratio.

Moreover, we evaluated our approach using three different word embedding models on seven different benchmark datasets. Our quantitative and qualitative analyses show that the performance of the system depends on the size and domain of the dataset. Ideally, these language models are trained on in-domain texts. Luckily, this is quite feasible as the models do not require annotated data. Normalising the datasets and entity mention queries may also boost coverage.

Not all issues can be resolved by carefully tuning the experiment parameters. The fact that often the approach does suggest a relevant entity type within the first 10 types also presents interesting avenues of research for post-processing.

One of the goals of our research is to discover more information about entities that are not present in a given knowledge base or for which the knowledge base does not contain sufficient relevant information to reason with. An entity typing step can provide us with likely entity types, a subsequent relation extraction may be used to further rerank the entity types. For non-NIL entities, we can also investigate whether having a fine grained entity type available may help improve entity linking.

Overall, our method provides a promising first step in using implicit and explicit domain knowledge for entity typing.

Acknowledgements. The research for this paper was made possible by the CLARIAH-CORE project financed by NWO: http://www.clariah.nl.

References

1. ACE (Automatic Content Extraction) english annotation guidelines for entities (2006). http://www.ldc.upenn.edu/Projects/ACE/
2. Bengio, Y., Ducharme, R., Vincent, P., Jauvin, C.: A neural probabilistic language model. J. Mach. Learn. Res. **3**, 1137–1155 (2003)
3. Cano, A.E., Rizzo, G., Varga, A., Rowe, M., Stankovic, M., Dadzie, A.S.: Making sense of microposts (#Microposts2014) named entity extraction & linking challenge. In: 4th International Workshop on Making Sense of Microposts. #Microposts (2014)

4. Elsner, M., Charniak, E., Johnson, M.: Structured generative models for unsupervised named-entity clustering. In: Proceedings of Human Language Technologies: The 2009 Annual Conference of the North American Chapter of the Association for Computational Linguistics (NAACL 2009), pp. 164–172 (2009)
5. van Erp, M., Ilievski, F., Rospocher, M., Vossen, P.: Missing Mr. Brown and buying an Abraham Lincoln - dark entities and DBpedia. In: Proceedings of NLP & DBpedia 2015 Workshop in Conjunction with 14th International Semantic Web Conference (ISWC 2015). CEUR Workshop Proceedings (2015)
6. van Erp, M., Mendes, P., Paulheim, H., Ilievski, F., Plu, J., Rizzo, G., Waitelonis, J.: Evaluating entity linking: an analysis of current benchmark datasets and a roadmap for doing a better job. In: Proceedings of LREC 2016 (2016). Preprint available from: https://mariekevanerp.files.wordpress.com/2012/06/evaluating-entity-linking-1.pdf
7. Grishman, R., Sundheim, B.M.: Message understanding conference - 6: a brief history. In: Proceedings International Conference on Computational Linguistics (1996)
8. Hachey, B., Radford, W., Nothman, J., Honnibal, M., Curran, J.R.: Evaluating entity linking with Wikipedia. Artif. Intell. **9**, 130–150 (2013)
9. Hoffart, J., Yosef, M.A., Bordin, I., Fürstenau, H., Pinkal, M., Spaniol, M., Taneva, B., Thater, S., Weikum, G.: Robust disambiguation of named entities. In: Conference on Empirical Methods in Natural Language Processing. EMNLP (2011)
10. Kittur, A., Chi, E.H., Suh, B.: What's in Wikipedia?: mapping topics and conflict using socially annotated category structure. In: Proceedings of the SIGCHI Conference on Human Factors in Computing Systems (CHI 2009), pp. 1509–1512. ACM, New York (2009)
11. Lewis, D.D., Yang, Y., Rose, T.G., Li, F.: RCV1: a new benchmark collection for text categorization research. J. Mach. Learn. Res. **5**, 361–397 (2004)
12. Mendes, P.N., Jakob, M., García-Silva, A., Bizer, C.: Dbpedia spotlight: shedding light on the web of documents. In: Proceedings of the 7th International Conference on Semantic Systems (I-SEMANTICS 2011), Graz, Austria. ACM New York, 7–9 September 2011
13. Mihalcea, R., Csomai, A.: Wikify! linking document to encyclopedic knowledge. In: Proceedings of the Sixteenth ACM Conference on Information and Knowledge Management (CIKM 2007), pp. 233–242 (2007)
14. Mikolov, T., Chen, K., Corrado, G., Dean, J.: Efficient estimation of word representations in vector space. In: arXiv preprint arXiv:1301.3781 (2013)
15. Mikolov, T., Sutskever, I., Chen, K., Corrado, G., Dean, J.: Distributed representations of words and phrases and their compositionality. In: Proceedings of NIPS (2013)
16. Milne, D., Witten, I.H.: Learning to link with Wikipedia. In: Proceedings of the 17th ACM Conference on Information and Knowledge Management (CIKM 2008), pp. 509–518 (2008)
17. Minard, A.L., Speranza, M., Urizar, R., na Altuna, B., van Erp, M., Schoen, A., van Son, C.: MEANTIME, the newsreader multilingual event and time corpus. In: Proceedings of the 10th Edition of the Language Resources and Evaluation Conference (LREC 2016) (2016)
18. Nadeau, D.: Semi-supervised named entity recognition: learning to recognize 100 entity types with little supervision. Ph.D. thesis, University of Ottawa (2007)
19. Nguyen, T.H., Grishman, R.: Relation extraction: perspective from convolutional neural networks. In: Proceedings of NAACL-HLT 2015, Denver, Colorado, USA, pp. 39–48, 31 May – 5 June 2015

20. Nuzzolese, A.G., Gentile, A.L., Presutti, V., Gangemi, A., Garigliotti, D., Navigli, R.: Open knowledge extraction challenge. In: Gandon, F., Cabrio, E., Stankovic, M., Zimmermann, A. (eds.) SemWebEval 2015. CCIS, vol. 548, pp. 3–15. Springer, Cham (2015). doi:10.1007/978-3-319-25518-7_1

21. Rizzo, G., Cano Amparo, E., Pereira, B., Varga, A.: Making sense of microposts (#Microposts2015) named entity recognition & linking challenge. In: 5th International Workshop on Making Sense of Microposts. #Microposts (2015)

22. Rizzo, G., Troncy, R.: NERD: a framework for unifying named entity recognition and disambiguation extraction tools. In: 13th Conference of the European Chapter of the Association for computational Linguistics (EACL 2012) (2012)

23. Röder, M., Usbeck, R., Hellmann, S., Gerber, D., Both, A.: N3-a collection of datasets for named entity recognition and disambiguation in the NLP interchange format. In: 9th Language Resources and Evaluation Conference. LREC (2014)

24. Sang, E.F.T.K.: Introduction to the CoNLL-2002 shared task: language-independent named entity recognition. In: Proceedings of CoNLL-2002, Taipei, Taiwan (2002)

25. Sekine, S., Sudo, K., Nobata, C.: Extended named entity hierarchy. In: Proceedings of the Third International Conference on Language Resources and Evaluation, pp. 1818–1824 (2002)

26. Sienčnik, S.K.: Adapting word2vec to named entity recognition. In: Proceedings of the 20th Nordic Conference of Computational Linguistics (NODALIDA 2015), Vilnius, Lithuania, pp. 239–243, 11–13 May 2015

27. Tjong Kim Sang, E.F., De Meulder, F.: Introduction to the CoNLL-2003 shared task: language-independent named entity recognition. In: Conference on Computational Natural Language Learning. CoNLL (2003)

28. Usbeck, R., Ngomo, A.C.N., Röder, M., Gerber, D., Coelho, S.A., Auer, S., Both, A.: AGDISTIS -graph-based disambiguation of named entities using linked data. In: Proceedings of the 13th International Semantic Web Conference (ISWC 2014), Riva del Garda, Italy, pp. 457–471, October 2014

29. Waitelonis, J., Exeler, C., Sack, H.: Linked data enabled generalized vector space model to improve document retrieval. In: Proceedings of NLP & DBpedia 2015 Workshop in Conjunction with 14th International Semantic Web Conference (ISWC2015). CEUR Workshop Proceedings (2015)

WC3: Analyzing the Style of Metadata Annotation Among Wikipedia Articles by Using Wikipedia Category and the DBpedia Metadata Database

Masaharu Yoshioka[✉]

Graduate School of Information Science and Technology, Hokkaido University,
N14 W9, Kita-ku, Sapporo 060-0814, Japan
yoshioka@ist.hokudai.ac.jp

Abstract. WC3 (Wikipedia Category Consistency Checker) is a system
that supports the analysis of the metadata-annotation style in Wikipedia
articles belonging to a particular Wikipedia category (the subcategory of
"Categories by parameter") by using the DBpedia metadata database.
This system aims to construct an appropriate SPARQL query to represent the category and compares the retrieved results and articles that
belong to the category. In this paper, we introduce WC3 and extend the
algorithm to analyze efficiently additional varieties of Wikipedia category. We also discuss the metadata-annotation quality of the Wikipedia
by using WC3. URL of WC3 is http://wnews.ist.hokudai.ac.jp/wc3/ and
related files are available at http://wnews.ist.hokudai.ac.jp/wc3/KEKI.

1 Introduction

Wikipedia[1] is a free, Wiki-based encyclopedia that covers a wide variety of topics.
Particularly for articles about named entities (e.g., person or artifact), metadata (e.g., writer or birthplace) about those entities are usually organized in
"infoboxes" displayed at the start of the articles. By extracting these items of
information, the DBpedia database [1] has been constructed. Because it covers
a wide variety of information about named entities, DBpedia has been used as
a core element of Linked Open Data [2] and for semantic annotation [3].

Another important source of information about metadata is Wikipedia category. For example, YAGO2 [4] extracts type information from them. In the
Wikipedia category structure, groups of categories can have an ancestor category such as "Categories by parameter". Most categories are then represented
in a set-and-topic style (e.g., "Cities in France"), whereby an original set (e.g.,
"cities") is divided into smaller topic categories according to a parameter value
(e.g., "France"). However, because of failures in the DBpedia metadata extraction and/or incomplete coverage in assigning appropriate Wikipedia categories
to the articles, there are some articles whose metadata obtained by DBpedia are

---

[1] http://www.wikipedia.org/.

© Springer International Publishing AG 2017
M. van Erp et al. (Eds.): ISWC 2016 Workshops, LNCS 10579, pp. 119–136, 2017.
https://doi.org/10.1007/978-3-319-68723-0_10

inconsistent with the metadata information based on the Wikipedia category structure.

To analyze the differences between these two information resources, we previously proposed WC3 (WC-triple: Wikipedia Category Consistency Checker)[2] based on the DBpedia metadata database [5]. This system aims to construct an appropriate SPARQL query to represent a Wikipedia category that is a subcategory of "Categories by parameter". A comparison between the queried Wikipedia articles and articles that belong to the category identifies articles that lack appropriate metadata annotation and articles that are candidates for category assignment. WC3 was a first attempt to analyze Wikipedia categories in systematic approach and can find out many inconsistent metadata annotation and/or Wikipedia category labels in the Wikipedia, mostly based on human errors and misunderstanding of the metadata annotation or Wikipedia category definition described as natural language text in the category description pages.

However, the system had a scalability problem and a lack of flexibility in constructing the SPARQL queries. In this paper, we propose an extended algorithm that uses sample articles and regular expressions for efficient and flexible SPARQL query construction based on the simple analysis of Wikipedia category strings. System performance and the consistency of the metadata annotation in Wikipedia are also addressed by applying the system to large numbers of Wikipedia categories.

The paper proceeds as follows. In Sect. 2, we briefly review research on the quality of the Wikipedia and DBpedia and support tools for enhancing Wikipedia contents. Section 3 introduces WC3 and proposes an extended algorithm for producing efficient and flexible SPARQL queries. In Sect. 4, we describe the Wikipedia category structure related to "Categories by parameter" and discuss the results of an analysis by WC3. Section 5 concludes.

2   Related Work

There have been several approaches to analyzing the quality of the Wikipedia and the DBpedia. The first approach was to check the contents of Wikipedia articles manually. Giles *et al.* [6] conducted an expert comparison between the scientific contents of Wikipedia and Encyclopedia Britannica, finding that Wikipedia had almost the same accuracy and quality as Encyclopedia Britannica. Another approach was an evaluation based on the editorial history of Wikipedia articles. Stvilla *et al.* [7] proposed a framework for information-quality assessment and confirmed that the quality of the Wikipedia can measure based on article edit history metadata, such as edit histories, discussions and vote logs. Kittur *et al.* [8] pointed out that it is important to maintain good coordination among editors to improve the quality of Wikipedia articles. Hu *et al.* [9] developed a quality measurement model for Wikipedia articles based on the quality of the editors.

---

[2] The link for WC3 has moved from http://wnews.ist.hokudai.ac.jp/wc3/ to http://wnews.ist.hokudai.ac.jp/wc3/old.

With respect to the quality of the DBpedia, a common approach is to compare metadata obtained from different information resources. Mendes *et al.* [10] implemented a system, "Sieve", that supports assessment of the linked-data quality. This work found that the coverage of DBpedia in different languages depends on the contents (e.g., the Portuguese DBpedia has more information about Brazilian municipalities than the English DBpedia) and a framework for integrating this information was proposed. Yoshioka and Kando [11] proposed a framework for an automatic method of discovering links between entries in GeoNames and Wikipedia articles. During this automatic link-discovery process, the system uncovered many errors about coordinate information in Wikipedia. Another approach was based on the revision history of Wikipedia. Orlandi and Passant [12] analyzed DBpedia according to provenance information based on the revision history [12].

This research on analyzing the quality of the Wikipedia and the DBpedia did not focus on how to support Wikipedia editors in improving the quality of Wikipedia articles. One approach to quality improvement on the academic side is entity linking. Mihalcea and Csomai [13] proposed an automatic keyword-extraction system based on Wikipedia articles. This system supported the addition of links to corresponding Wikipedia articles based on the extracted results. There have been several attempts to utilize this framework (e.g., link discovery in the English Wikipedia [14] and cross-language link discovery in Wikipedia [15]). For Wikipedia category maintenance, PetScan[3] is a simple tool for identifying candidate Wikipedia articles by manually constructing queries based on information about Wikipedia articles such as templates, links, and Wikidata. However, it is not easy to construct appropriate queries for analyzing the Wikipedia categories manually. In contrast, WC3 [5] supports the automatic construction of candidate queries based on DBpedia information.

Torres *et al.* [16] proposed a framework for selecting representative Wikipedia category paths by using DBpedia SPARQL queries and Wikipedia category information. However, this system did not aim to support Wikipedia's volunteer editors.

3 WC3

3.1 Prototype of WC3

WC3 aims to support Wikipedia's volunteer editors by checking the consistency of metadata annotation related to a given Wikipedia category [5]. This is achieved by constructing an appropriate SPARQL query to represent a set-and-topic-style category. This query is used to retrieve results from DBpedia, with the retrieved results being compared with articles that belong to the given category to check their metadata-annotation consistency.

This system analyzes all metadata for articles that belong to the target category and identifies candidate attributes for use in the SPARQL query.

---

[3] https://petscan.wmflabs.org/.

There are two types of candidate attributes. First, there are attributes that involve the topic-related restriction. These are selected by applying an F-measure (harmonic mean of precision and recall) to articles belonging to the target category. Second, there are attributes that involve the set-related restriction, which are selected by using articles that belong to the target category (e.g., "Song written by Paul McCartney") or sibling categories of the target (e.g., "Song written by Bob Dylan"). Finally, the system checks all combinations of these candidate attributes and uses a combination with the highest F-measure to generate the SPARQL query.

A prototype version of WC3 was implemented by using the 2014 version of DBpedia data[4] as the resource description framework (RDF) database and a Wikipedia dump database (dated 2014/11/6) for finding sibling categories.

3.2   Issues with the WC3 Prototype

The prototype system can generate appropriate SPARQL queries for the Wikipedia categories whose topic is person names (e.g., "Songs written by ..." or "Films directed by ..."). However, it fails to generate appropriate queries for categories whose topic is not represented as a simple metadata value (e.g., articles belonging to "People from Tokyo" could have the metadata "Tokyo", "Tokyo, Japan", or "Bunkyo-ku, Tokyo" for "birthplace").

Moreover, because the prototype system tries to use the metadata of all articles belonging to a category, it requires excessive time to analyze a category with many articles (more than 10 min for a category with 1,000 articles). To achieve better usability, the performance should be improved.

3.3   Proposal for a New Algorithm

To address the problems in the prototype system, we propose a new algorithm for WC3 by adopting the following approaches.

- Use of a FILTER function in SPARQL:
  To generate appropriate SPARQL queries for categories whose topic is not represented as a simple metadata value, the FILTER function finds related topic attributes to use for constructing the SPARQL query. To identify topic-related strings, sibling categories are used to exclude the shared string. For example, "1981 births" has sibling categories such as "1982 births" and "1972 births". The topic-related string "1981" is extracted by excluding "births", and such strings are used in generating a query via the FILTER function. Since this operation is language independent, this method can be applied analysis of DBpedia and Wikipedia in any languages.
- Generating SPARQL queries by combining topic and set restrictions:
  In the prototype system, articles in sibling categories are used to identify set-related restrictions. However, there is a computational cost and the

---

[4] http://wiki.dbpedia.org/Datasets2014/.

quality of the result relies on randomly selected sibling categories. In this paper, metadata that have a type predicate (rdf:type[5]) or a short description (dbp:shortDescription) are selected for set-related restrictions. The final SPARQL query is generated by the highest-ranked combination of one of these set-related restrictions and a topic-related restriction, using an F-measure ranking derived from an analysis of articles belonging to the category.

– Reduction of the response time by reducing the number of SPARQL queries to an RDF database. This involves:

  • Using sample articles for categories with many articles:
    Because candidate attributes that have higher recall should have a higher recall for the sample articles, we use sample articles to identify candidate attributes for generating an appropriate query. A page size threshold (pst) parameter is introduced to control the size of the sampling.

  • Exclusion of candidate attributes whose recall is low:
    Because the recall of a candidate attribute is an upper bound on that of a combined-query attribute, these attributes are not used in the final query, making it unnecessary to calculate the precision and F-measure for such attributes. Therefore, we sort candidate attributes based on their recall values and use mca (maximum candidate attributes) to pick mca set-related attributes and mca topic-related attributes for restriction.

– Classification of errors into related subcategories and other subcategories:
  When the system analyzes a category that has articles and subcategories, there are several cases where the constructed query retrieves articles that belong to a subcategory (e.g., the SPARQL query for "People from Tokyo" would retrieve articles belonging to "Writers from Tokyo", which is a subcategory of "People from Tokyo"). To clarify the difference between errors related to subcategories and other errors, error articles are checked as to whether they belong to subcategory.

Figure 1 shows basic computational flow to construct SPARQL queries for the Wikipedia category.

Generation of Candidate Strings for FILTER. All of the topic and set Wikipedia category have parent categories for characterizing topic and/or set (See example for "1991 births" in Fig. 1). For each parent category, child categories share common substring that represents the characteristics of the category (topic and/or set). Identification of common topic or set by using common shared substings is useful to identify topic or set keywords of the target category. From the topic related parent category (e.g., "1991" for "1991 births"), topic related keywords are identified by shared common strings ("1991"). The system can generate set based keywords "births" by removing the shared common string.

---

[5] In this paper, we use the abbreviations "dbo", "dbp", "dbr" and "rdf", for "http://dbpedia.org/ontology", "http://dbpedia.org/property", "http://dbpedia.org/resource" and "http://www.w3.org/1999/02/22-rdf-syntax-ns#type".

Generation of candidate strings for FILTER (e.g., 1991_births)

1. **Selection of sibling categories** (5 random sibling categories for each parent category)
   e.g., "1992_births", "1993_births", … for parent category "1990s_births"
   "1991_death", "1991_events", … for parent category "1991".
2. **Identification of shared substring from the beginning and end of the category name.**
   e.g., "199" and "_births" for "1990s_births" and "1991_" for "1991".
3. **Generation of candidate strings by removing shared strings from the beginning and end.**
   e.g., "1_births", "1991" and "1" for removing shared sub string "199", "_births", and both of them.
   "births" for removing shared string "1991"

Generation of Restriction Candidates for SPARQL query

1. **Selection of articles belongs to the category**
   When number of the articles for the category is larger than *Pst* (50 in this experiment), *Pst* articles are selected by using limit operation of SPARQL. For other case, all article is used as selected one.
2. **Generation of candidate restriction for each selected article**
   The system retrieves all metadata of each article and classify metadata whose predicates are rdf:type or dbp:shortDescription for set-related restriction (*Src*) and others for topic-related restriction (*Trc*). When the objects of the metadata contains candidate substrings for FILTER, system also counts number of these metadata (*SrFc*, *TrFc*).

A'dia Mathies <dbr:A'dia Mathies rdf:type http://xmlns.com/foaf/0.1/Person> <dbr:A'dia Mathies rdf:type dbo:Person> <dbr:A'dia Mathies dbo:birthDate 1991-03-18> …	Set-restriction rdf:type http://xmlns.com/foaf/0.1/Person rdf:type dbo:Person … Topic-restriction dbo:birthDate 1991-03-18 dbo:birthDate ?o1 FILTER regex (?o1, "1991") dbo:birthDate ?o1 FILTER regex (?o1, "91") …

3. **Selection of candidate restrictions**
   Calculate F-measure of each restriction and select top *mca* (10 in this experiment) candidates from *Trc*, *Src,*, *SrFc*, and *TrFc*.
4. **Generation of combined restrictions**
   Calculate F-measure of all combinations of one from set-related restrictions (*Src*, *SrFc*) and one from topic-related restrictions (*Trc*, *TrFc*).

Combined-restriction as SPARQL for "1991_births"
?s rdf:type http://xmlns.com/foaf/0.1/Person . ?s dbo:birthDate ?o1 . FILTER regex (?o1, "1991")
?s rdf:type dbo:Person . ?s dbo:birthDate ?o1 . FILTER regex (?o1, "1991")
?s rdf:type http://xmlns.com/foaf/0.1/Person . ?s dbo:birthDate ?o1 FILTER regex (?o1, "91")
?s rdf:type dbo:Person . ?s dbo:birthDate ?o1 . FILTER regex (?o1, "91")
…

System output

Output is classified based on the comparison between retrieved results and category information

	Retrieved by query	Not retrieved
Articles belongs to the category	Found	NotFound
Articles belongs to the subcategory	Error(Subcategory)	
Others	Error(Other)	

Fig. 1. Basic computational flow to constructed SPARQL queries for the Wikipedia category

Topic related keywords can also be extracted by using set related parent category (e.g., "1990s birth").

In order to identify these strings, the system retrieves sibling categories (at most 20 categories) for each parent category and check most common shared string from the beginning and from the end (e.g., "1991", "199", "births" for "1991 births"). Substrings generated by removing these common shared strings are used as candidate strings for FILTER (e.g., "births", "1 births", "1991", "1").

Generation of Restriction Candidates for SPARQL Query. At first, the system selects at most *pst* articles from the articles belongs to the category. Metadata[6] associated with selected articles are used for generating restriction candidates. At first metadata are classified into two by checking predicates. Metadata whose predicate is rdf:type or dbp:shortDescription are used for set-related restriction set (Src) and others are used for topic-related restriction set (Trc). In addition, for each restriction set, metadata whose object contains candidate FILTER strings are used for restriction sets with FILTER ($SrFc, TrFc$).

From these candidate restrictions sets, the system calculates F-measure for those candidates and selects top mca restrictions for each restriction set for generating combination restrictions ($Src_{mca}.Trc_{mca}.SrFc_{mca}.TrFc_{mca}$).

All pairs of set-related and topic-related restrictions (one from Src_{mca} or $SrFc_{mca}$ and the other from Trc_{mca} or $TrFc_{mca}$) are evaluated by using F-measure. Restriction with highest F-measure is selected as a final result of the system.

The system presents results for related articles categorized into the following four types.

Found: articles that belong to the category and are retrieved by the query.

NotFound: articles that belong to the category but cannot be retrieved by the query.

Error(SubCategory): articles that belong to a subcategory of the target category but are retrieved by the query.

Error(Other): articles that do not belong to the category or its subcategories but are retrieved by the query.

In addition, the system can also accept a manually constructed SPARQL query for evaluating the Wikipedia category. If the user is not satisfied by the generated query, the user can modify the SPARQL query to evaluate the category. A common example of this function is to use automatic query generation for a sibling category and then to modify the topic-related restriction.

---

[6] Metadata related to YAGO [4] are excluded, because it uses Wikipedia category information as a resource to extract the data.

4 Wikipedia Category Analysis Using WC3

4.1 Candidate Wikipedia Categories

For this analysis, we used DBpedia (2015-04 version)[7] and the Wikipedia dump database used for DBpedia (dated 2015/04/03)[8].

WC3 aims to analyze set-and-topic-style Wikipedia categories. In the Wikipedia category structure, those categories are located as subcategories of "Categories by parameter". However, not all subcategories of "Categories by parameter" are set-and-topic-style Wikipedia categories. For example, Wikipedia category "Hokkaido University" is a subcategory of "Universities and colleges in Sapporo", but this category is a topic category and it is difficult to analyze by using WC3. Because most of these topic categories have articles with the same name and most set-and-topic-style categories do not have such articles, we exclude categories that have articles with the same name from candidates of set-and-topic-style categories.

To discuss the applicability of WC3, we now consider the proportion of categories found as set-and-topic-style categories from subcategories of "Categories by parameter". The total number of Wikipedia categories that have at least one category page or subcategory is 1,251,889. In addition, there are a number of categories classified as "stubs". Because "stubs" categories indicate that articles are under development, it is inappropriate to analyze them by using WC3-like tools. After excluding such "stubs" categories, Wikipedia has 1,238,392 categories in total. From these categories, we found 691,671 set-and-topic-style categories and 565,431 categories with at least one article in such a category. From these results, we estimate that WC3 may help about half (565,431/1,238,392 = 45.7%) of the categories in Wikipedia.

4.2 System Implementation and Examples

We have implemented WC3 (FILTER version)[9], which is an extended version of WC3. The parameter values used for the system were $pst = 50$ and $mca = 10$.

Figure 2 shows a screenshot of WC3 (FILTER version) and an example of the system output for the "1973 births" category.

The system constructed the following SPARQL query and found 9,259, 1,342 and 322 articles categorized as Found, NotFound, and Error, respectively (recall = 9,259/(9,259 + 1,342) = 0.87 and precision = 9,259/(9,259 + 322) = 0.97).

```
SELECT DISTINCT ?s
WHERE {
?s rdf:type http://xmlns.com/foaf/0.1/Person .
?s dbo:birthDate ?o1 .
```

---

[7] http://wiki.dbpedia.org/Downloads2015-04.

[8] Files related to the experiments are available at http://wnews.ist.hokudai.ac.jp/wc3/KEKI.

[9] http://wnews.ist.hokudai.ac.jp/wc3/.

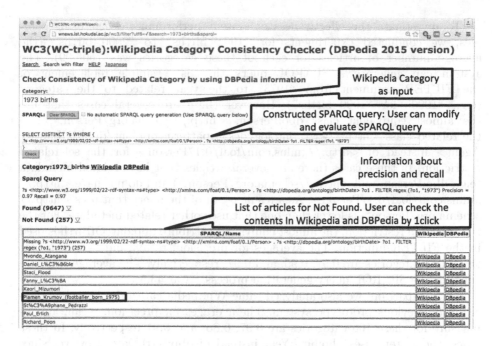

Fig. 2. Screenshot of WC3 (FILTER version)

```
FILTER regex (?o1, ''1973'')
}
MINUS { ?s dbo:wikiPageRedirects ?o . }}
```

When checking the NotFound articles, there were two problematic cases. The first arises from a lack of information from the infobox (e.g., "Bai Xiaoyun"[10]). The second arises from inconsistent annotations for the infobox and Wikipedia categories (e.g., "Plamen Krumov (footballer, born 1975)"). Most of the Error articles are candidates for the category except articles that have multiple instances of birthDate information. There are two types of such articles. First, articles can involve more than one person (e.g., "List of Playboy Playmates of 1994"). Second, there can be problems with the metadata extraction in DBpedia (e.g., "Anjali Sudhakar" and "Barbara Kanam"). For the latter case, most of those articles have another birth-related Wikipedia category (e.g., "1972 births" for "Anjali Sudhakar" and "1970 births" for "Barbara Kanam"), but there are cases where the annotated Wikipedia category seems to be wrong ("Anjali Sudhakar") and the DBpedia metadata seems to be wrong ("Barbara Kanam"). In addition, there are many articles whose metadata are inconsistent with Wikipedia categories (e.g., "Ratish Nanda" being categorized as "1974 births").

---

[10] All English Wikipedia articles referred to in this section were accessed on March 7, 2017.

To investigate the effectiveness of system performance when using samples, we conducted experiments for "1901 births" to "2000 births" (the average number of articles for each of these categories was 7606.99). The system can generate a SPARQL query that is a simple replacement of the FILTER argument from "1973" to the year related to the category (e.g., "1900" for "1900 births"). However there are several cases that use the almost equivalent class dbo:Person, <http://schema.org/Person>, and the related class <http://www.ontologydesignpatterns.org/ont/dul/DUL.owl#Agent> instead of <http://xmlns.com/foaf/0.1/Person> for the set-related restriction. In these cases there are several articles that have related metadata (e.g., dbo:Person, and <http://schema.org/Person>) and do not have <http://xmlns.com/foaf/0.1/Person>. As a result recall of the query that uses <http://schema.org/Person> is lower than one that uses other related metadata. For the topic-related restriction, the almost equivalent attribute dbp:dateOfBirth is used in the FILTER function. The averages for precision, recall, and F-measure for these 100 categories are 0.97, 0.96, and 0.97, respectively. The average response time for the SPARQL query-generation process[11] was 43.5 s.

For comparison, we also analyzed same Wikipedia categories by using algorithm of previous prototype system. The averages for precision, recall, and F-measure for these 100 categories are 0.97, 0.86, and 0.91 respectively. In these cases, the system uses dbo:birthYear instead of dbp:birthDate. However since there are several articles that have dbp:birthDate and do not have dbp:birthYear, recall of the previous system is worse than one of the proposed system. The average response time for the SPARQL query-generation was increased to 849 s because of checking all metadata of the articles belongs to the category. Using sample articles significantly reduce the time of SPARQL query generation for Wikipedia categories with large number of articles.

To evaluate the quality of the generated SPARQL queries, we applied WC3 to the sample of set-and-topic-style Wikipedia categories. We sorted the candidate set-and-topic-style categories based on the number of articles belonging to the category that were not redirects and used the top 5,000 for evaluation. To investigate the appropriateness of using FILTER in the analysis, we conducted experiments that compared the proposed system with a non-FILTER system. Since non-Filter system cannot select candidates from restrictions with FILTER, the system selects $2 * mca(20)$ candidates from set-related restriction set (Src) and topic-related restriction set (Trc) for combination generation.

For this experiment, because articles belonging to the Error(SubCategory) may be candidate articles when the subcategories are not divided into more detailed subcategories, $prec' = |PP|/(|PA| - |PS|)$, where PP, PA, and PS represent articles that satisfy the SPARQL query in the Wikipedia category, all Wikipedia data, and in subcategories of the Wikipedia category, respectively.

Table 1 is cross tables of recall and precision for the SPARQL queries constructed by the proposed system. The averages of precision, recall, and F-measure

---

[11] Because the rendering time for displaying the results was less than a second, the time for SPARQL query generation is almost equivalent to the total response time.

Table 1. Cross table of recall and precision for SPARQL queries constructed by the proposed system

Precision recall	[0, 0.2)	[0.2, 0.4)	[0.4, 0.6)	[0.6, 0.8)	[0.8, 1]	Total
[0, 0.2)	621	271	171	76	45	1184
[0.2, 0.4)	156	208	183	70	33	650
[0.4, 0.6)	123	167	172	126	46	634
[0.6, 0.8)	125	135	116	98	51	525
[0.8, 1)	153	188	270	416	980	2007
Total	1178	969	912	786	1155	5000

for the proposed system were 0.50, 0.58, and 0.48 respectively and ones for the proposed system were 0.46, 0.54, and 0.44 respectively. We can confirm that there is a statistically significant difference between these two systems in terms of the Wilcoxon's signed rank test ($p < 0.01$).

Since there are many categories that FILTER function is not useful, queries for 2,280 categories are same for the both system. For 2,096 categories, the proposed system uses FILTER function in the queries and F-measure of the proposed system is better than the non-FILTER system in 1,779 cases; average difference of precision, recall and F-measure between both system for these queries are 0.12, 0.10 and 0.11 respectively. This result suggests that there are many Wikipedia categories where FILTER function is useful to represent appropriate queries. On the contrary, non-FILTER system generates better queries for 45 categories by using restriction candidates from Src and trc that are not used in the proposed system (non-FILTER system can use 20 candidates instead of 10 for Src and Trc). Average difference of precision, recall and F-measure between both system for these queries are, 0.06, 0.00, and 0.03 respectively.

Since it takes too much computational time to generate SPARQL query results for the Wikipedia category with large numbers of articles by using previous system, we select last 100 categories from this 5,000 tested categories for analyzing the difference between proposed system and the previous system. Number of the articles for these categories are 309 to 313 (average is 310.82). Average response time for these categories of the proposed system and previous system were 28.22 and 28.78 s respectively. From this results, we found that computational time of the proposed system doesn't affect so much by the number of the articles for the target category, but the computational time of the previous system improves well due to the decreasing number of articles. One of the reason why the system requires more time is a number of combined candidates checked for the best combination. The proposed system checks 400 candidates (2 * mca * 2 * mca) that is larger than previous system. In addition, SPARQL queries with FILTER function tend to take longer time than queries without FILTER function.

For the retrieval performance, average of the precision, recall and F-measure of the proposed system 0.45, 0.57, 0.43, respectively and ones for the previous system are 0.43, 0.55 and 0.42, respectively and there is no statistically significant difference. Out of 100 categories, the proposed system is better than previous system in 37 categories and the previous system is better in 25 categories. Most of the queries for former categories cases use FILTER function. For the latter categories cases, the previous system generate SPARQL queries that uses combination of two set-related restrictions or two topic-related restrictions. Appropriateness of using those queries are discussed in success and failure analysis (Sect. 4.3).

4.3   Success and Failure Analysis

From these analysis, there are many Wikipedia categories that requires FILTER function to represent more appropriate SPARQL queries for representing Wikipedia categories. However, the performance of the system is inadequate for analyzing all of those Wikipedia categories. In analyzing the success and failure of the system, the following examples are used in the discussion.

- Higher recall and precision:
  "Films directed by D. W. Griffith" (prec = 0.99, recall = 0.95) is an example of using a combination of set-related and topic-related restrictions.

```
SELECT DISTINCT ?s
WHERE {
?s rdf:type dbo:Wikidata:Q11424 .
?s dbo:director ?o1 .
FILTER regex (?o1, ''D._W._Griffith'')
MINUS { ?s dbo:wikiPageRedirects ?o . }}
```

  "Portugal international footballers"(prec = 0.99, recall = 0.97) is an example of when using a corresponding property to represent topic-related restriction is good enough.

```
SELECT DISTINCT ?s
WHERE {
?s rdf:type http://xmlns.com/foaf/0.1/Person .
?s dbp:nationalteam dbr:Portugal_national_football_team .
MINUS { ?s dbo:wikiPageRedirects ?o . }}
```

  In both cases, Error(Other) contains candidate articles for adding the target Wikipedia category.
- Higher precision with modest recall:
  "University of Michigan alumni" (prec = 0.98, recall = 0.31) is an example of using a combination of set-related and topic-related restrictions.

```
SELECT DISTINCT ?s
WHERE {
?s rdf:type http://xmlns.com/foaf/0.1/Person .
?s dbp:almaMater ?o1 .
FILTER regex (?o1,"University_of_Michigan")
MINUS { ?s dbo:wikiPageRedirects ?o . }}
```

"Hungarian canoeist" (prec = 1.0, recall = 0.67) is an example of using a corresponding shortDescription if one exists. Recall of the query relies on the one for extracting such metadata from the articles.

```
SELECT DISTINCT ?s
WHERE {
?s <dbp:shortDescription Hungarian canoeist> .
?s <http://purl.org/dc/elements/1.1/description Hungarian canoeist> .
MINUS { ?s dbo:wikiPageRedirects ?o . }}
```

For "University of Michigan alumni", Error(Other) contains candidates for adding the target. In both cases, NotFound contains a list of articles without common metadata annotation.

– Higher recall with modest precision:
"American metalcore musical groups" (prec = 0.44 and recall = 0.75) is an example of requiring additional topic-related restrictions.

```
SELECT DISTINCT ?s
WHERE {
?s rdf:type dbo:Band .
?s dbo:genre dbr:Metalcore
MINUS { ?s dbo:wikiPageRedirects ?o . }}
```

"Public high schools in North Carolina" (prec = 0.42 and recall = 0.78) is an example of lacking an appropriate set-related restriction.

```
SELECT DISTINCT ?s
WHERE {
?s rdf:type dbo:School .
?s dbo:city ?o1 .
FILTER regex (?o1,"North_Carolina")
MINUS { ?s dbo:wikiPageRedirects ?o . }}
```

In the former case, precision would be improved by adding another restriction to represent "American".

– Modest recall and modest precision:
"People from Tokyo" (prec = 0.53, recall = 0.59) is an example that shows incomplete coverage of articles for a particular Wikipedia category.

```
SELECT DISTINCT ?s
WHERE {
?s rdf:type http://xmlns.com/foaf/0.1/Person .
?s dbp:placeOfBirth ?o1 .
FILTER regex (?o1, ''Tokyo'')
MINUS { ?s dbo:wikiPageRedirects ?o . }}
```

Because "People from Tokyo" is a category for "people who were born in or who are residents of Tokyo, Japan,"[12] the generated SPARQL query seems to be reasonable for selecting people who were born in Tokyo. Low precision means the incompleteness of the Wikipedia category as an index for articles. Therefore, Error(Other) may contain candidates for adding the target. However, this query does not find people who were residents of Tokyo.

– Modest precision and lower recall:
"Diesel locomotives of the United States" (prec = 0.6, recall = 0.38) is an example that case.

```
SELECT DISTINCT ?s
WHERE {
?s rdf:type dbo:Locomotive .
?s dbo:powerType dbr:Diesel-electric_transmission .
MINUS { ?s dbo:wikiPageRedirects ?o . }}
```

There are many categories that requires two or more set-related or topic-related restriction by using the DBpedia metadata. For example, this case requires set-related restriction locomotive that uses diesel-electric transmission and one topic-related restriction "American". However, this query only represents set-related restriction. There are many cases that a topic-related restriction is used for representing a set-related restriction for the category that requires two or more set-related restrictions are necessary. For this case, since most of the machines that uses diesel-electric transmission are locomotive, removing such set-related restriction improves the retrieval performance. The previous system can generate query with one restriction, following query is generated by the previous system (prec = 0.60, recall = 0.51),

```
SELECT DISTINCT ?s
WHERE {
?s dbo:powerType dbr:Diesel-electric_transmission .
MINUS { ?s dbo:wikiPageRedirects ?o . }}
```

– Lower recall and precision:
"American Professional Soccer League players" (prec = 0.17, recall = 0.22) is an example that case.

```
SELECT DISTINCT ?s
WHERE {
?s rdf:type dbo:SoccerManager
?s dbp:birthPlace dbr:United_States
MINUS { ?s dbo:wikiPageRedirects ?o . }}
```

There are several cases for which the vocabulary of DBpedia or extracted metadata are insufficient to represent set or topic. In such a case, the system

---

[12] https://en.wikipedia.org/wiki/Category:People_from_Tokyo.

tries to find out related metadata (for this case, since many soccer players become soccer managers, metadata related to the soccer managers is used) from the candidates. Compare to the previous system, the proposed system only select from set-related metadata. On the contrary, since there is no such restriction in the previous system, following query generated by the previous system has better performance (prec = 0.37, recall = 0.20). It may be related to that "American Professional Soccer League" was founded in 1990.

```
SELECT DISTINCT ?s
WHERE {
?s dbp:years 1990
?s dbp:birthPlace dbr:United_States
MINUS { ?s dbo:wikiPageRedirects ?o . }}
```

Another type of failure relates to the limitations of the constructed SPARQL queries (e.g., "20th-century births" and "S.League players"). For the category "20th-century births", it might be better to use the regular expression "19??" for dateOfBirth, but this system cannot generate such an expression. For "S.League players", it is necessary to represent the relationship between "S.League" and "team" in constructing an appropriate query.

4.4   Discussion

Issues Related to the Set-Related Restrictions and Topic-Related Restrictions. From this analysis, we can confirm that the new WC3 (FILTER version) can construct appropriate SPARQL queries when there are appropriate metadata for set-related and topic-related restriction. However, there are several cases that queries constructed as a combination of two topic-related restriction or two set-related restriction. It is necessary to discuss the effect of using set-related or topic-related restriction queries only from the viewpoint of improving the performance and improving the readability to the user.

Trade Off Between Computational Cost and Variations of Candidate SPARQL Queries. In addition, the use of sample articles reduced the response time for the categories with large numbers of articles. However, due to the number of candidates generated for finding appropriate query and using FILTER function requires more computation time in general. It is necessary to discuss the trade-off between the computational time and the final performance.

Support for Constructing More Flexible SPARQL Queries. As discussed using the example of lower recall and precision, it is necessary to adopt another strategy for constructing SPARQL queries involving the use of regular expression patterns for particular types of topic (e.g., "19??" for "20th-century") and the use of a part-whole relationship (e.g., relationships between "team" and "league" or between "region" and "subregion").

In addition to such automatic support, we also plan to make a database of SPARQL queries for representing Wikipedia category that were confirmed manually. The voluntary editors can use such query when it exists. Even if there is no corresponding query for the category, query of sibling categories that have different topic-related restriction may be helpful to construct new SPARQL query by replacing topic-related restriction(s) in the query.

Feedback for DBpedia and Wikipedia. As discussed using the example of "1973 births", there are several cases where the extracted DBpedia metadata themselves are inconsistent (e.g., multiple birthDates for one person). It would be better to have a framework for identifying such problems, which would improve the quality of metadata extraction in DBpedia.

One of the problems of the system is using fixed DBpedia data. It is helpful to discover problems that exist at that time. However, an editor cannot check if a particular update would adequately address the problem. For such problems, it would be preferable to have a framework for updating the DBpedia database based on editor-supplied updates. For example, when a volunteer editor checks a Wikipedia category, the editor could request updates for the metadata of articles that belong to that category. We have already started to discuss this issue with the Japanese DBpedia community[13] for the Japanese version of WC3, called WC3ja[14].

5   Conclusions

In this paper, we have proposed an extension of WC3 that uses a FILTER function to represent SPARQL queries for Wikipedia categories whose common metadata values are not just simple values. In addition, we have investigated a sampling approach for identifying candidate attributes and confirmed that sampling approach may improve response time when there are large number of articles belongs to the category. For the retrieval performance, there are many cases that usage of FILTER function improves the retrieval performance, but there are also many cases that combination of a set-related restriction and a topic-related restriction is not good enough compared to the two set-related restrictions or two topic-related restrictions. It is necessary to discuss the effectiveness of using those categories from the viewpoint of retrieval performance and readability.

This approach has also clarified the situation where editors aiming to maintain an infobox may overlook the addition of appropriate Wikipedia categories. As a result, there are several Wikipedia categories whose coverage of related articles is incomplete.

Even though there remain several issues with the new system, providing the system to Wikipedia's volunteer editors would help improve the quality of

---

[13] http://ja.dbpedia.org/.
[14] http://wnews.ist.hokudai.ac.jp/wc3ja.

metadata annotation in Wikipedia and, as a result, the quality of the DBpedia would also improve.

Acknowledgment. This work was partially supported by JSPS KAKENHI Grant Number 25280035 and 16H01756.

References

1. Bizer, C., Lehmann, J., Kobilarov, G., Auer, S., Becker, C., Cyganiak, R., Hellmann, S.: DBpedia - a crystallization point for the web of data. Web Seman. Sci. Serv. Agents World Wide Web **7**, 154–165 (2009)
2. Bizer, C., Heath, T., Berners-Lee, T.: Linked data - the story so far. Int. J. Seman. Web Inf. Syst. **5**, 1–22 (2009)
3. Mendes, P.N., Jakob, M., García-Silva, A., Bizer, C.: DBpedia spotlight: shedding light on the web of documents. In: Proceedings of the 7th International Conference on Semantic Systems. I-Semantics 2011, pp. 1–8. ACM, New York (2011)
4. Hoffart, J., Suchanek, F.M., Berberich, K., Weikum, G.: YAGO2: a spatially and temporally enhanced knowledge base from Wikipedia. Artif. Intell. **194**, 28–61 (2013)
5. Yoshioka, M., Loban, R.: WC3: Wikipedia category consistency checker based on DBPedia. In: Proceedings of 11th International Conference on Signal-Image Technology & Internet-Based Systems, pp. 712–718 (2015)
6. Giles, J.: Internet encyclopaedias go head to head. Nature **438**, 900–901 (2005)
7. Stvilia, B., Gasser, L., Twidale, M.B., Smith, L.C.: A framework for information quality assessment. J. Am. Soc. Inform. Sci. Technol. **58**, 1720–1733 (2007)
8. Kittur, A., Kraut, R.E.: Harnessing the wisdom of crowds in Wikipedia: quality through coordination. In: Proceedings of the 2008 ACM Conference on Computer Supported Cooperative Work, CSCW 2008, pp. 37–46. ACM, New York (2008)
9. Hu, M., Lim, E.P., Sun, A., Lauw, H.W., Vuong, B.Q.: Measuring article quality in Wikipedia: models and evaluation. In: Proceedings of the Sixteenth ACM Conference on Information and Knowledge Management, CIKM 2007, pp. 243–252. ACM, New York (2007)
10. Mendes, P.N., Mühleisen, H., Bizer, C.: Sieve: linked data quality assessment and fusion. In: Proceedings of the 2012 Joint EDBT/ICDT Workshops, EDBT-ICDT 2012, pp. 116–123. ACM, New York (2012)
11. Yoshioka, M., Kando, N.: Issues for linking geographical open data of GeoNames and Wikipedia. In: Takeda, H., Qu, Y., Mizoguchi, R., Kitamura, Y. (eds.) JIST 2012. LNCS, vol. 7774, pp. 375–381. Springer, Heidelberg (2013). doi:10.1007/978-3-642-37996-3_32
12. Orlandi, F., Passant, A.: Modelling provenance of DBpedia resources using Wikipedia contributions. Web Seman. Sci. Serv. Agents World Wide Web **9**, 149–164 (2011). Provenance in the Semantic Web
13. Mihalcea, R., Csomai, A.: Wikify!: linking documents to encyclopedic knowledge. In: Proceedings of the Sixteenth ACM Conference on Information and Knowledge Management, CIKM 2007, pp. 233–242. ACM, New York (2007)
14. Xu, M., Wang, Z., Bie, R., Li, J., Zheng, C., Ke, W., Zhou, M.: Discovering missing semantic relations between entities in Wikipedia. In: Alani, H., et al. (eds.) ISWC 2013. LNCS, vol. 8218, pp. 673–686. Springer, Heidelberg (2013). doi:10.1007/978-3-642-41335-3_42

136   M. Yoshioka

15. Tang, L.X., Kang, I.S., Kimura, F., Lee, Y.H., Trotman, A., Geva, S., Xu, Y.: Overview of the NTCIR-10 cross-lingual link discovery task. In: Proceedings of the 10th NTCIR Workshop Meeting on Evaluation of Information Access Technologies: Information Retrieval, Quesiton Answering, and Cross-Lingual Information Access, pp. 8–38 (2013)
16. Torres, D., Molli, P., Skaf-Molli, H., Diaz, A.: Improving Wikipedia with DBpedia. In: Proceedings of the 21st International Conference on World Wide Web. WWW 2012 Companion, pp. 1107–1112. ACM, New York (2012)

Author Index

Printed in the United States
By Bookmasters